Springer Texts in Statistics

Advisors:
George Casella   Stephen Fienberg   Ingram Olkin

Springer Texts in Statistics

(continued after index)

Larry Wasserman

All of Nonparametric Statistics

With 52 Illustrations

 Springer

Larry Wasserman
Department of Statistics
Carnegie Mellon University
Pittsburgh, PA 15213-3890
USA
larry@stat.cmu.edu

ISBN 978-1-4419-2044-7   e-ISBN 978-0-387-30623-0

Printed on acid-free paper.

Printed in the United States of America.   (MVY)

9 8 7 6 5 4 3 2 1

springeronline.com

To Isa

Preface

There are many books on various aspects of nonparametric inference such as density estimation, nonparametric regression, bootstrapping, and wavelets methods. But it is hard to find all these topics covered in one place. The goal of this text is to provide readers with a single book where they can find a brief account of many of the modern topics in nonparametric inference.

The book is aimed at master's-level or Ph.D.-level statistics and computer science students. It is also suitable for researchers in statistics, machine learning and data mining who want to get up to speed quickly on modern nonparametric methods. My goal is to quickly acquaint the reader with the basic concepts in many areas rather than tackling any one topic in great detail. In the interest of covering a wide range of topics, while keeping the book short, I have opted to omit most proofs. Bibliographic remarks point the reader to references that contain further details. Of course, I have had to choose topics to include and to omit, the title notwithstanding. For the most part, I decided to omit topics that are too big to cover in one chapter. For example, I do not cover classification or nonparametric Bayesian inference.

The book developed from my lecture notes for a half-semester (20 hours) course populated mainly by master's-level students. For Ph.D.-level students, the instructor may want to cover some of the material in more depth and require the students to fill in proofs of some of the theorems. Throughout, I have attempted to follow one basic principle: never give an estimator without giving a confidence set.

The book has a mixture of methods and theory. The material is meant to complement more method-oriented texts such as Hastie et al. (2001) and Ruppert et al. (2003).

After the Introduction in Chapter 1, Chapters 2 and 3 cover topics related to the empirical CDF such as the nonparametric delta method and the bootstrap. Chapters 4 to 6 cover basic smoothing methods. Chapters 7 to 9 have a higher theoretical content and are more demanding. The theory in Chapter 7 lays the foundation for the orthogonal function methods in Chapters 8 and 9. Chapter 10 surveys some of the omitted topics.

I assume that the reader has had a course in mathematical statistics such as Casella and Berger (2002) or Wasserman (2004). In particular, I assume that the following concepts are familiar to the reader: distribution functions, convergence in probability, convergence in distribution, almost sure convergence, likelihood functions, maximum likelihood, confidence intervals, the delta method, bias, mean squared error, and Bayes estimators. These background concepts are reviewed briefly in Chapter 1.

Data sets and code can be found at:

www.stat.cmu.edu/~larry/all-of-nonpar

I need to make some disclaimers. First, the topics in this book fall under the rubric of "modern nonparametrics." The omission of traditional methods such as rank tests and so on is not intended to belittle their importance. Second, I make heavy use of large-sample methods. This is partly because I think that statistics is, largely, most successful and useful in large-sample situations, and partly because it is often easier to construct large-sample, nonparametric methods. The reader should be aware that large-sample methods can, of course, go awry when used without appropriate caution.

I would like to thank the following people for providing feedback and suggestions: Larry Brown, Ed George, John Lafferty, Feng Liang, Catherine Loader, Jiayang Sun, and Rob Tibshirani. Special thanks to some readers who provided very detailed comments: Taeryon Choi, Nils Hjort, Woncheol Jang, Chris Jones, Javier Rojo, David Scott, and one anonymous reader. Thanks also go to my colleague Chris Genovese for lots of advice and for writing the LATEX macros for the layout of the book. I am indebted to John Kimmel, who has been supportive and helpful and did not rebel against the crazy title. Finally, thanks to my wife Isabella Verdinelli for suggestions that improved the book and for her love and support.

Larry Wasserman
Pittsburgh, Pennsylvania
July 2005

Contents

1
Introduction

In this chapter we briefly describe the types of problems with which we will be concerned. Then we define some notation and review some basic concepts from probability theory and statistical inference.

1.1 What Is Nonparametric Inference?

The basic idea of nonparametric inference is to use data to infer an unknown quantity while making as few assumptions as possible. Usually, this means using statistical models that are infinite-dimensional. Indeed, a better name for nonparametric inference might be infinite-dimensional inference. But it is difficult to give a precise definition of nonparametric inference, and if I did venture to give one, no doubt I would be barraged with dissenting opinions.

For the purposes of this book, we will use the phrase nonparametric inference to refer to a set of modern statistical methods that aim to keep the number of underlying assumptions as weak as possible. Specifically, we will consider the following problems:

1. (Estimating the distribution function). Given an IID sample $X_1, \ldots, X_n \sim F$, estimate the CDF $F(x) = \mathbb{P}(X \leq x)$. (Chapter 2.)

2. (Estimating functionals). Given an IID sample $X_1, \ldots, X_n \sim F$, estimate a functional $T(F)$ such as the mean $T(F) = \int x \, dF(x)$. (Chapters 2 and 3.)

3. (Density estimation). Given an IID sample $X_1, \ldots, X_n \sim F$, estimate the density $f(x) = F'(x)$. (Chapters 4, 6 and 8.)

4. (Nonparametric regression or curve estimation). Given $(X_1, Y_1), \ldots, (X_n, Y_n)$ estimate the regression function $r(x) = \mathbb{E}(Y|X = x)$. (Chapters 4, 5, 8 and 9.)

5. (Normal means). Given $Y_i \sim N(\theta_i, \sigma^2)$, $i = 1, \ldots, n$, estimate $\theta = (\theta_1, \ldots, \theta_n)$. This apparently simple problem turns out to be very complex and provides a unifying basis for much of nonparametric inference. (Chapter 7.)

In addition, we will discuss some unifying theoretical principles in Chapter 7. We consider a few miscellaneous problems in Chapter 10, such as measurement error, inverse problems and testing.

Typically, we will assume that distribution F (or density f or regression function r) lies in some large set $\mathfrak{F}$ called a **statistical model**. For example, when estimating a density f, we might assume that

$$f \in \mathfrak{F} = \left\{ g : \int (g''(x))^2 dx \leq c^2 \right\}$$

which is the set of densities that are not "too wiggly."

1.2   Notation and Background

Here is a summary of some useful notation and background. See also Table 1.1.

Let $a(x)$ be a function of x and let F be a cumulative distribution function. If F is absolutely continuous, let f denote its density. If F is discrete, let f denote instead its probability mass function. The mean of a is

$$\mathbb{E}(a(X)) = \int a(x) dF(x) \equiv \begin{cases} \int a(x) f(x) dx & \text{continuous case} \\ \sum_j a(x_j) f(x_j) & \text{discrete case.} \end{cases}$$

Let $\mathbb{V}(X) = \mathbb{E}(X - \mathbb{E}(X))^2$ denote the variance of a random variable. If $X_1, \ldots, X_n$ are n observations, then $\int a(x) d\widehat{F}_n(x) = n^{-1} \sum_i a(X_i)$ where $\widehat{F}_n$ is the **empirical distribution** that puts mass $1/n$ at each observation X_i.

Symbol	Definition
$x_n = o(a_n)$	$\lim_{n \to \infty} x_n/a_n = 0$
$x_n = O(a_n)$	$|x_n/a_n|$ is bounded for all large n
$a_n \sim b_n$	$a_n/b_n \to 1$ as $n \to \infty$
$a_n \asymp b_n$	a_n/b_n and b_n/a_n are bounded for all large n
$X_n \rightsquigarrow X$	convergence in distribution
$X_n \xrightarrow{\text{P}} X$	convergence in probability
$X_n \xrightarrow{\text{a.s.}} X$	almost sure convergence
$\widehat{\theta}_n$	estimator of parameter θ
bias	$\mathbb{E}(\widehat{\theta}_n) - \theta$
se	$\sqrt{\mathbb{V}(\widehat{\theta}_n)}$ (standard error)
$\widehat{\text{se}}$	estimated standard error
MSE	$\mathbb{E}(\widehat{\theta}_n - \theta)^2$ (mean squared error)
Φ	CDF of a standard Normal random variable
z_α	$\Phi^{-1}(1 - \alpha)$

TABLE 1.1. Some useful notation.

Brief Review of Probability. The **sample space** Ω is the set of possible outcomes of an experiment. Subsets of Ω are called **events**. A class of events $\mathcal{A}$ is called a σ-**field** if (i) $\emptyset \in \mathcal{A}$, (ii) $A \in \mathcal{A}$ implies that $A^c \in \mathcal{A}$ and (iii) $A_1, A_2, \ldots, \in \mathcal{A}$ implies that $\bigcup_{i=1}^{\infty} A_i \in \mathcal{A}$. A **probability measure** is a function $\mathbb{P}$ defined on a σ-field $\mathcal{A}$ such that $\mathbb{P}(A) \geq 0$ for all $A \in \mathcal{A}$, $\mathbb{P}(\Omega) = 1$ and if $A_1, A_2, \ldots \in \mathcal{A}$ are disjoint then

$$\mathbb{P}\left(\bigcup_{i=1}^{\infty} A_i\right) = \sum_{i=1}^{\infty} \mathbb{P}(A_i).$$

The triple $(\Omega, \mathcal{A}, \mathbb{P})$ is called a **probability space**. A **random variable** is a map $X : \Omega \to \mathbb{R}$ such that, for every real x, $\{\omega \in \Omega : X(\omega) \leq x\} \in \mathcal{A}$.

A sequence of random variables X_n **converges in distribution** (or converges weakly) to a random variable X, written $X_n \rightsquigarrow X$, if

$$\mathbb{P}(X_n \leq x) \to \mathbb{P}(X \leq x) \tag{1.1}$$

as $n \to \infty$, at all points x at which the CDF

$$F(x) = \mathbb{P}(X \leq x) \tag{1.2}$$

is continuous. A sequence of random variables X_n **converges in probability** to a random variable X, written $X_n \xrightarrow{\text{P}} X$, if,

$$\text{for every } \epsilon > 0, \quad \mathbb{P}(|X_n - X| > \epsilon) \to 0 \quad \text{as } n \to \infty. \tag{1.3}$$

A sequence of random variables X_n **converges almost surely** to a random variable X, written $X_n \xrightarrow{\text{a.s.}} X$, if

$$\mathbb{P}(\lim_{n \to \infty} |X_n - X| = 0) = 1. \tag{1.4}$$

The following implications hold:

$$X_n \xrightarrow{\text{a.s.}} X \quad \text{implies that} \quad X_n \xrightarrow{\text{P}} X \quad \text{implies that} \quad X_n \rightsquigarrow X. \tag{1.5}$$

Let g be a continuous function. Then, according to the **continuous mapping theorem**,

$$\begin{aligned} X_n \rightsquigarrow X &\quad \text{implies that} \quad g(X_n) \rightsquigarrow g(X) \\ X_n \xrightarrow{\text{P}} X &\quad \text{implies that} \quad g(X_n) \xrightarrow{\text{P}} g(X) \\ X_n \xrightarrow{\text{a.s.}} X &\quad \text{implies that} \quad g(X_n) \xrightarrow{\text{a.s.}} g(X) \end{aligned}$$

According to **Slutsky's theorem**, if $X_n \rightsquigarrow X$ and $Y_n \rightsquigarrow c$ for some constant c, then $X_n + Y_n \rightsquigarrow X + c$ and $X_n Y_n \rightsquigarrow cX$.

Let $X_1, \ldots, X_n \sim F$ be IID. The **weak law of large numbers** says that if $\mathbb{E}|g(X_1)| < \infty$, then $n^{-1} \sum_{i=1}^{n} g(X_i) \xrightarrow{\text{P}} \mathbb{E}(g(X_1))$. The **strong law of large numbers** says that if $\mathbb{E}|g(X_1)| < \infty$, then $n^{-1} \sum_{i=1}^{n} g(X_i) \xrightarrow{\text{a.s.}} \mathbb{E}(g(X_1))$.

The random variable Z has a standard Normal distribution if it has density $\phi(z) = (2\pi)^{-1/2} e^{-z^2/2}$ and we write $Z \sim N(0,1)$. The CDF is denoted by $\Phi(z)$. The α upper quantile is denoted by z_α. Thus, if $Z \sim N(0,1)$, then $\mathbb{P}(Z > z_\alpha) = \alpha$.

If $\mathbb{E}(g^2(X_1)) < \infty$, the **central limit theorem** says that

$$\sqrt{n}(\overline{Y}_n - \mu) \rightsquigarrow N(0, \sigma^2) \tag{1.6}$$

where $Y_i = g(X_i)$, $\mu = \mathbb{E}(Y_1)$, $\overline{Y}_n = n^{-1} \sum_{i=1}^{n} Y_i$ and $\sigma^2 = \mathbb{V}(Y_1)$. In general, if

$$\frac{(X_n - \mu)}{\widehat{\sigma}_n} \rightsquigarrow N(0,1)$$

then we will write

$$X_n \approx N(\mu, \widehat{\sigma}_n^2). \tag{1.7}$$

According to the **delta method**, if g is differentiable at μ and $g'(\mu) \neq 0$ then

$$\sqrt{n}(X_n - \mu) \rightsquigarrow N(0, \sigma^2) \implies \sqrt{n}(g(X_n) - g(\mu)) \rightsquigarrow N(0, (g'(\mu))^2 \sigma^2). \tag{1.8}$$

A similar result holds in the vector case. Suppose that X_n is a sequence of random vectors such that $\sqrt{n}(X_n - \mu) \rightsquigarrow N(0, \Sigma)$, a multivariate, mean 0 normal with covariance matrix Σ. Let g be differentiable with gradient ∇g such that $\nabla_\mu \neq 0$ where ∇_μ is ∇g evaluated at μ. Then

$$\sqrt{n}(g(X_n) - g(\mu)) \rightsquigarrow N\left(0, \nabla_\mu^T \Sigma \nabla_\mu\right). \tag{1.9}$$

Statistical Concepts. Let $\mathfrak{F} = \{f(x; \theta) : \theta \in \Theta\}$ be a parametric model satisfying appropriate regularity conditions. The **likelihood function** based on IID observations $X_1, \ldots, X_n$ is

$$\mathcal{L}_n(\theta) = \prod_{i=1}^n f(X_i; \theta)$$

and the **log-likelihood function** is $\ell_n(\theta) = \log \mathcal{L}_n(\theta)$. The maximum likelihood estimator, or MLE $\widehat{\theta}_n$, is the value of θ that maximizes the likelihood. The **score function** is $s(X; \theta) = \partial \log f(x; \theta)/\partial\theta$. Under appropriate regularity conditions, the score function satisfies $\mathbb{E}_\theta(s(X; \theta)) = \int s(x; \theta)f(x; \theta)dx = 0$. Also,

$$\sqrt{n}(\widehat{\theta}_n - \theta) \rightsquigarrow N(0, \tau^2(\theta))$$

where $\tau^2(\theta) = 1/I(\theta)$ and

$$I(\theta) = \mathbb{V}_\theta(s(x; \theta)) = \mathbb{E}_\theta(s^2(x; \theta)) = -\mathbb{E}_\theta\left(\frac{\partial^2 \log f(x; \theta)}{\partial\theta^2}\right)$$

is the **Fisher information**. Also,

$$\frac{(\widehat{\theta}_n - \theta)}{\widehat{se}} \rightsquigarrow N(0, 1)$$

where $\widehat{se}^2 = 1/(nI(\widehat{\theta}_n))$. The Fisher information I_n from n observations satisfies $I_n(\theta) = nI(\theta)$; hence we may also write $\widehat{se}^2 = 1/(I_n(\widehat{\theta}_n))$.

The bias of an estimator $\widehat{\theta}_n$ is $\mathbb{E}(\widehat{\theta}) - \theta$ and the the mean squared error MSE is $\text{MSE} = \mathbb{E}(\widehat{\theta} - \theta)^2$. The **bias–variance decomposition** for the MSE of an estimator $\widehat{\theta}_n$ is

$$\text{MSE} = \text{bias}^2(\widehat{\theta}_n) + \mathbb{V}(\widehat{\theta}_n). \tag{1.10}$$

1.3   Confidence Sets

Much of nonparametric inference is devoted to finding an estimator $\widehat{\theta}_n$ of some quantity of interest θ. Here, for example, θ could be a mean, a density or a regression function. But we also want to provide confidence sets for these quantities. There are different types of confidence sets, as we now explain.

Let $\mathfrak{F}$ be a class of distribution functions F and let θ be some quantity of interest. Thus, θ might be F itself, or F' or the mean of F, and so on. Let C_n be a set of possible values of θ which depends on the data $X_1, \ldots, X_n$. To emphasize that probability statements depend on the underlying F we will sometimes write $\mathbb{P}_F$.

1.11 Definition. C_n *is a* **finite sample** $1 - \alpha$ **confidence set** *if*

$$\inf_{F \in \mathfrak{F}} \mathbb{P}_F(\theta \in C_n) \geq 1 - \alpha \quad \text{for all } n. \tag{1.12}$$

C_n *is a* **uniform asymptotic** $1 - \alpha$ **confidence set** *if*

$$\liminf_{n \to \infty} \inf_{F \in \mathfrak{F}} \mathbb{P}_F(\theta \in C_n) \geq 1 - \alpha. \tag{1.13}$$

C_n *is a* **pointwise asymptotic** $1 - \alpha$ **confidence set** *if,*

$$\text{for every } F \in \mathfrak{F}, \quad \liminf_{n \to \infty} \mathbb{P}_F(\theta \in C_n) \geq 1 - \alpha. \tag{1.14}$$

If $||\cdot||$ denotes some norm and $\widehat{f}_n$ is an estimate of f, then a **confidence ball** for f is a confidence set of the form

$$C_n = \left\{ f \in \mathfrak{F} : \ ||f - \widehat{f}_n|| \leq s_n \right\} \tag{1.15}$$

where s_n may depend on the data. Suppose that f is defined on a set $\mathcal{X}$. A pair of functions (ℓ, u) is a $1 - \alpha$ **confidence band** or **confidence envelope** if

$$\inf_{f \in \mathfrak{F}} \mathbb{P}\big(\ell(x) \leq f(x) \leq u(x) \text{ for all } x \in \mathcal{X}\big) \geq 1 - \alpha. \tag{1.16}$$

Confidence balls and bands can be finite sample, pointwise asymptotic and uniform asymptotic as above. When estimating a real-valued quantity instead of a function, C_n is just an interval and we call C_n a confidence interval.

Ideally, we would like to find finite sample confidence sets. When this is not possible, we try to construct uniform asymptotic confidence sets. The last resort is a pointwise asymptotic confidence interval. If C_n is a uniform asymptotic confidence set, then the following is true: for any $\delta > 0$ there exists an $n(\delta)$ such that the coverage of C_n is at least $1 - \alpha - \delta$ for all $n > n(\delta)$. With a pointwise asymptotic confidence set, there may not exist a finite $n(\delta)$. In this case, the sample size at which the confidence set has coverage close to $1 - \alpha$ will depend on f (which we don't know).

1.17 Example. Let $X_1, \ldots, X_n \sim$ Bernoulli(p). A pointwise asymptotic $1 - \alpha$ confidence interval for p is

$$\widehat{p}_n \pm z_{\alpha/2} \sqrt{\frac{\widehat{p}_n(1 - \widehat{p}_n)}{n}} \tag{1.18}$$

where $\widehat{p}_n = n^{-1} \sum_{i-1}^n X_i$ It follows from Hoeffding's inequality (1.24) that a finite sample confidence interval is

$$\widehat{p}_n \pm \sqrt{\frac{1}{2n} \log\left(\frac{2}{\alpha}\right)}. \quad \blacksquare \tag{1.19}$$

1.20 Example (Parametric models). Let

$$\mathfrak{F} = \{f(x; \theta) : \theta \in \Theta\}$$

be a parametric model with scalar parameter θ and let $\widehat{\theta}_n$ be the maximum likelihood estimator, the value of θ that maximizes the likelihood function

$$\mathcal{L}_n(\theta) = \prod_{i=1}^n f(X_i; \theta).$$

Recall that under suitable regularity assumptions,

$$\widehat{\theta}_n \approx N(\theta, \widehat{se}^2)$$

where

$$\widehat{se} = (I_n(\widehat{\theta}_n))^{-1/2}$$

is the estimated standard error of $\widehat{\theta}_n$ and $I_n(\theta)$ is the Fisher information. Then

$$\widehat{\theta}_n \pm z_{\alpha/2}\widehat{se}$$

is a pointwise asymptotic confidence interval. If $\tau = g(\theta)$ we can get an asymptotic confidence interval for τ using the delta method. The MLE for τ is $\widehat{\tau}_n = g(\widehat{\theta}_n)$. The estimated standard error for τ is $\widehat{se}(\widehat{\tau}_n) = \widehat{se}(\widehat{\theta}_n)|g'(\widehat{\theta}_n)|$. The confidence interval for τ is

$$\widehat{\tau}_n \pm z_{\alpha/2}\widehat{se}(\widehat{\tau}_n) = \widehat{\tau}_n \pm z_{\alpha/2}\widehat{se}(\widehat{\theta}_n)|g'(\widehat{\theta}_n)|.$$

Again, this is typically a pointwise asymptotic confidence interval. $\blacksquare$

1.4   Useful Inequalities

At various times in this book we will need to use certain inequalities. For reference purposes, a number of these inequalities are recorded here.

Markov's Inequality. Let X be a non-negative random variable and suppose that $\mathbb{E}(X)$ exists. For any $t > 0$,

$$\mathbb{P}(X > t) \leq \frac{\mathbb{E}(X)}{t}. \tag{1.21}$$

Chebyshev's Inequality. Let $\mu = \mathbb{E}(X)$ and $\sigma^2 = \mathbb{V}(X)$. Then,

$$\mathbb{P}(|X - \mu| \geq t) \leq \frac{\sigma^2}{t^2}. \tag{1.22}$$

Hoeffding's Inequality. Let $Y_1, \ldots, Y_n$ be independent observations such that $\mathbb{E}(Y_i) = 0$ and $a_i \leq Y_i \leq b_i$. Let $\epsilon > 0$. Then, for any $t > 0$,

$$\mathbb{P}\left(\sum_{i=1}^{n} Y_i \geq \epsilon\right) \leq e^{-t\epsilon} \prod_{i=1}^{n} e^{t^2(b_i - a_i)^2/8}. \tag{1.23}$$

Hoeffding's Inequality for Bernoulli Random Variables. Let $X_1, \ldots, X_n \sim \text{Bernoulli}(p)$. Then, for any $\epsilon > 0$,

$$\mathbb{P}\left(|\overline{X}_n - p| > \epsilon\right) \leq 2e^{-2n\epsilon^2} \tag{1.24}$$

where $\overline{X}_n = n^{-1} \sum_{i=1}^{n} X_i$.

Mill's Inequality. If $Z \sim N(0, 1)$ then, for any $t > 0$,

$$\mathbb{P}(|Z| > t) \leq \frac{2\phi(t)}{t} \tag{1.25}$$

where ϕ is the standard Normal density. In fact, for any $t > 0$,

$$\left(\frac{1}{t} - \frac{1}{t^3}\right)\phi(t) \; < \; \mathbb{P}(Z > t) \; < \; \frac{1}{t}\phi(t) \tag{1.26}$$

and

$$P(Z > t) < \frac{1}{2}e^{-t^2/2}. \tag{1.27}$$

Berry–Esséen Bound. Let $X_1, \ldots, X_n$ be IID with finite mean $\mu = \mathbb{E}(X_1)$, variance $\sigma^2 = \mathbb{V}(X_1)$ and third moment, $\mathbb{E}|X_1|^3 < \infty$. Let $Z_n = \sqrt{n}(\overline{X}_n - \mu)/\sigma$. Then

$$\sup_z |\mathbb{P}(Z_n \leq z) - \Phi(z)| \leq \frac{33}{4} \frac{\mathbb{E}|X_1 - \mu|^3}{\sqrt{n}\sigma^3}. \tag{1.28}$$

Bernstein's Inequality. Let $X_1, \ldots, X_n$ be independent, zero mean random variables such that $-M \leq X_i \leq M$. Then

$$\mathbb{P}\left(\left| \sum_{i=1}^n X_i \right| > t \right) \leq 2 \exp \left\{ -\frac{1}{2} \left(\frac{t^2}{v + Mt/3} \right) \right\} \tag{1.29}$$

where $v \geq \sum_{i=1}^n \mathbb{V}(X_i)$.

Bernstein's Inequality (Moment version). Let $X_1, \ldots, X_n$ be independent, zero mean random variables such that

$$\mathbb{E}|X_i|^m \leq \frac{m! M^{m-2} v_i}{2}$$

for all $m \geq 2$ and some constants M and v_i. Then,

$$\mathbb{P}\left(\left| \sum_{i=1}^n X_i \right| > t \right) \leq 2 \exp \left\{ -\frac{1}{2} \left(\frac{t^2}{v + Mt} \right) \right\} \tag{1.30}$$

where $v = \sum_{i=1}^n v_i$.

Cauchy–Schwartz Inequality. If X and Y have finite variances then

$$\mathbb{E}|XY| \leq \sqrt{\mathbb{E}(X^2)\mathbb{E}(Y^2)}. \tag{1.31}$$

Recall that a function g is **convex** if for each x, y and each $\alpha \in [0, 1]$,

$$g(\alpha x + (1 - \alpha)y) \leq \alpha g(x) + (1 - \alpha)g(y).$$

If g is twice differentiable, then convexity reduces to checking that $g''(x) \geq 0$ for all x. It can be shown that if g is convex then it lies above any line that touches g at some point, called a tangent line. A function g is **concave** if $-g$ is convex. Examples of convex functions are $g(x) = x^2$ and $g(x) = e^x$. Examples of concave functions are $g(x) = -x^2$ and $g(x) = \log x$.

Jensen's inequality. If g is convex then

$$\mathbb{E}g(X) \geq g(\mathbb{E}X). \tag{1.32}$$

If g is concave then

$$\mathbb{E}g(X) \leq g(\mathbb{E}X). \tag{1.33}$$

1.5 Bibliographic Remarks

References on probability inequalities and their use in statistics and pattern recognition include Devroye et al. (1996) and van der Vaart and Wellner (1996). To review basic probability and mathematical statistics, I recommend Casella and Berger (2002), van der Vaart (1998) and Wasserman (2004).

1.6 Exercises

1. Consider Example 1.17. Prove that (1.18) is a pointwise asymptotic confidence interval. Prove that (1.19) is a uniform confidence interval.

2. (Computer experiment). Compare the coverage and length of (1.18) and (1.19) by simulation. Take $p = 0.2$ and use $\alpha = .05$. Try various sample sizes n. How large must n be before the pointwise interval has accurate coverage? How do the lengths of the two intervals compare when this sample size is reached?

3. Let $X_1, \ldots, X_n \sim N(\mu, 1)$. Let $C_n = \overline{X}_n \pm z_{\alpha/2}/\sqrt{n}$. Is C_n a finite sample, pointwise asymptotic, or uniform asymptotic confidence set for μ?

4. Let $X_1, \ldots, X_n \sim N(\mu, \sigma^2)$. Let $C_n = \overline{X}_n \pm z_{\alpha/2} S_n/\sqrt{n}$ where $S_n^2 = \sum_{i=1}^{n} (X_i - \overline{X}_n)^2/(n-1)$. Is C_n a finite sample, pointwise asymptotic, or uniform asymptotic confidence set for μ?

5. Let $X_1, \ldots, X_n \sim F$ and let $\mu = \int x \, dF(x)$ be the mean. Let

$$C_n = \left(\overline{X}_n - z_{\alpha/2}\widehat{se}, \ \overline{X}_n + z_{\alpha/2}\widehat{se} \right)$$

where $\widehat{se}^2 = S_n^2/n$ and

$$S_n^2 = \frac{1}{n} \sum_{i=1}^{n} (X_i - \overline{X}_n)^2.$$

(a) Assuming that the mean exists, show that C_n is a $1 - \alpha$ pointwise asymptotic confidence interval.

(b) Show that C_n is not a uniform asymptotic confidence interval. *Hint*: Let $a_n \to \infty$ and $\epsilon_n \to 0$ and let $G_n = (1 - \epsilon_n)F + \epsilon_n \delta_n$ where δ_n is a pointmass at a_n. Argue that, with very high probability, for a_n large and ϵ_n small, $\int x \, dG_n(x)$ is large but $\overline{X}_n + z_{\alpha/2}\widehat{\mathrm{se}}$ is not large.

(c) Suppose that $\mathbb{P}(|X_i| \leq B) = 1$ where B is a known constant. Use Bernstein's inequality (1.29) to construct a finite sample confidence interval for μ.

2
Estimating the CDF and Statistical Functionals

The first problem we consider is estimating the CDF. By itself, this is not a very interesting problem. However, it is the first step towards solving more important problems such as estimating statistical functionals.

2.1 The CDF

We begin with the problem of estimating a CDF (cumulative distribution function). Let $X_1, \ldots, X_n \sim F$ where $F(x) = \mathbb{P}(X \leq x)$ is a distribution function on the real line. We estimate F with the empirical distribution function.

2.1 Definition. *The **empirical distribution function** $\widehat{F}_n$ is the* CDF *that puts mass $1/n$ at each data point X_i. Formally,*

$$\widehat{F}_n(x) = \frac{1}{n} \sum_{i=1}^{n} I(X_i \leq x) \tag{2.2}$$

where

$$I(X_i \leq x) = \begin{cases} 1 & \text{if } X_i \leq x \\ 0 & \text{if } X_i > x. \end{cases}$$

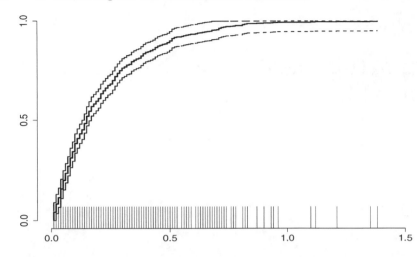

FIGURE 2.1. Nerve data. Each vertical line represents one data point. The solid line is the empirical distribution function. The lines above and below the middle line are a 95 percent confidence band.

2.3 Example (Nerve data). Cox and Lewis (1966) reported 799 waiting times between successive pulses along a nerve fiber. Figure 2.1 shows the data and the empirical CDF $\widehat{F}_n$. ∎

The following theorem gives some properties of $\widehat{F}_n(x)$.

2.4 Theorem. *Let $X_1, \ldots, X_n \sim F$ and let $\widehat{F}_n$ be the empirical* CDF. *Then:*

1. *At any fixed value of x,*

$$\mathbb{E}\left(\widehat{F}_n(x)\right) = F(x) \quad \text{and} \quad \mathbb{V}\left(\widehat{F}_n(x)\right) = \frac{F(x)(1 - F(x))}{n}.$$

Thus, MSE $= \frac{F(x)(1-F(x))}{n} \to 0$ *and hence* $\widehat{F}_n(x) \overset{\text{P}}{\longrightarrow} F(x)$.

2. (Glivenko–Cantelli Theorem).

$$\sup_x |\widehat{F}_n(x) - F(x)| \overset{\text{a.s.}}{\longrightarrow} 0.$$

3. (Dvoretzky–Kiefer–Wolfowitz (DKW) inequality). *For any $\epsilon > 0$,*

$$\mathbb{P}\left(\sup_x |F(x) - \widehat{F}_n(x)| > \epsilon\right) \le 2e^{-2n\epsilon^2}. \tag{2.5}$$

From the DKW inequality, we can construct a confidence set. Let $\epsilon_n^2 = \log(2/\alpha)/(2n)$, $L(x) = \max\{\widehat{F}_n(x) - \epsilon_n, 0\}$ and $U(x) = \min\{\widehat{F}_n(x) + \epsilon_n, 1\}$. It follows from (2.5) that for any F,

$$\mathbb{P}(L(x) \leq F(x) \leq U(x) \text{ for all } x) \geq 1 - \alpha.$$

Thus, $(L(x), U(x))$ is a nonparametric $1 - \alpha$ **confidence band**.[1]
To summarize:

2.6 Theorem. *Let*

$$
\begin{aligned}
L(x) &= \max\{\widehat{F}_n(x) - \epsilon_n, 0\} \\
U(x) &= \min\{\widehat{F}_n(x) + \epsilon_n, 1\}
\end{aligned}
$$

where

$$\epsilon_n = \sqrt{\frac{1}{2n} \log\left(\frac{2}{\alpha}\right)}.$$

Then, for all F and all n,

$$\mathbb{P}\left(L(x) \leq F(x) \leq U(x) \text{ for all } x\right) \geq 1 - \alpha.$$

2.7 Example. The dashed lines in Figure 2.1 give a 95 percent confidence band using $\epsilon_n = \sqrt{\frac{1}{2n} \log\left(\frac{2}{.05}\right)} = .048$. ∎

2.2   Estimating Statistical Functionals

A **statistical functional** $T(F)$ is any function of F. Examples are the mean $\mu = \int x \, dF(x)$, the variance $\sigma^2 = \int (x - \mu)^2 dF(x)$ and the median $m = F^{-1}(1/2)$.

2.8 Definition. *The* **plug-in estimator** *of $\theta = T(F)$ is defined by*

$$\widehat{\theta}_n = T(\widehat{F}_n). \tag{2.9}$$

A functional of the form $\int a(x) dF(x)$ is called a **linear functional**. Recall that $\int a(x) dF(x)$ is defined to be $\int a(x) f(x) dx$ in the continuous case

---

[1] There exist tighter confidence bands but we use the DKW band because it is simple.

and $\sum_j a(x_j)f(x_j)$ in the discrete case. The empirical CDF $\widehat{F}_n(x)$ is discrete, putting mass $1/n$ at each X_i. Hence, if $T(F) = \int a(x)dF(x)$ is a linear functional then we have:

The plug-in estimator for linear functional $T(F) = \int a(x)dF(x)$ is:

$$T(\widehat{F}_n) = \int a(x)d\widehat{F}_n(x) = \frac{1}{n}\sum_{i=1}^{n}a(X_i). \qquad (2.10)$$

Sometimes we can find the estimated standard error $\widehat{se}$ of $T(\widehat{F}_n)$ by doing some direct calculations. However, in other cases it is not obvious how to estimate the standard error. Later we will discuss methods for finding $\widehat{se}$. For now, let us just assume that somehow we can find $\widehat{se}$. In many cases, it turns out that

$$T(\widehat{F}_n) \approx N(T(F), \widehat{se}^2). \qquad (2.11)$$

In that case, an approximate $1 - \alpha$ confidence interval for $T(F)$ is then

$$T(\widehat{F}_n) \pm z_{\alpha/2}\,\widehat{se} \qquad (2.12)$$

where z_α is defined by: $\mathbb{P}(Z > z_\alpha) = \alpha$ with $Z \sim N(0,1)$. We will call (2.12) the **Normal-based interval.**

2.13 Example (The mean). Let $\mu = T(F) = \int x\,dF(x)$. The plug-in estimator is $\widehat{\mu} = \int x\,d\widehat{F}_n(x) = \overline{X}_n$. The standard error is $se = \sqrt{\mathbb{V}(\overline{X}_n)} = \sigma/\sqrt{n}$. If $\widehat{\sigma}$ denotes an estimate of σ, then the estimated standard error is $\widehat{se} = \widehat{\sigma}/\sqrt{n}$. A Normal-based confidence interval for μ is $\overline{X}_n \pm z_{\alpha/2}\,\widehat{\sigma}/\sqrt{n}$. ∎

2.14 Example (The variance). Let $\sigma^2 = \mathbb{V}(X) = \int x^2\,dF(x) - \left(\int x\,dF(x)\right)^2$. The plug-in estimator is

$$\begin{aligned}
\widehat{\sigma}^2 &= \int x^2 d\widehat{F}_n(x) - \left(\int xd\widehat{F}_n(x)\right)^2 \\
&= \frac{1}{n}\sum_{i=1}^{n}X_i^2 - \left(\frac{1}{n}\sum_{i=1}^{n}X_i\right)^2 \\
&= \frac{1}{n}\sum_{i=1}^{n}(X_i - \overline{X}_n)^2.
\end{aligned}$$

This is different than the usual unbiased sample variance

$$S_n^2 = \frac{1}{n-1}\sum_{i=1}^{n}(X_i - \overline{X}_n)^2.$$

In practice, there is little difference between $\widehat{\sigma}^2$ and S_n^2. ∎

2.15 Example (The skewness). Let μ and σ^2 denote the mean and variance of a random variable X. The skewness — which measures the lack of symmetry of a distribution — is defined to be

$$\kappa = \frac{\mathbb{E}(X - \mu)^3}{\sigma^3} = \frac{\int (x - \mu)^3 dF(x)}{\left\{ \int (x - \mu)^2 dF(x) \right\}^{3/2}}.$$

To find the plug-in estimate, first recall that $\widehat{\mu} = n^{-1} \sum_{i=1}^{n} X_i$ and $\widehat{\sigma}^2 = n^{-1} \sum_{i=1}^{n} (X_i - \widehat{\mu})^2$. The plug-in estimate of κ is

$$\widehat{\kappa} = \frac{\int (x - \mu)^3 d\widehat{F}_n(x)}{\left\{ \int (x - \mu)^2 d\widehat{F}_n(x) \right\}^{3/2}} = \frac{\frac{1}{n} \sum_{i=1}^{n} (X_i - \widehat{\mu})^3}{\widehat{\sigma}^3}. \quad ∎$$

2.16 Example (Correlation). Let $Z = (X, Y)$ and let $\rho = T(F) = \mathbb{E}(X - \mu_X)(Y - \mu_Y)/(\sigma_x \sigma_y)$ denote the correlation between X and Y, where $F(x, y)$ is bivariate. We can write $T(F) = a(T_1(F), T_2(F), T_3(F), T_4(F), T_5(F))$ where

$$T_1(F) = \int x \, dF(z) \quad T_2(F) = \int y \, dF(z) \quad T_3(F) = \int xy \, dF(z)$$
$$T_4(F) = \int x^2 \, dF(z) \quad T_5(F) = \int y^2 \, dF(z)$$

and

$$a(t_1, \ldots, t_5) = \frac{t_3 - t_1 t_2}{\sqrt{(t_4 - t_1^2)(t_5 - t_2^2)}}.$$

Replace F with $\widehat{F}_n$ in $T_1(F)$, ..., $T_5(F)$, and take

$$\widehat{\rho} = a(T_1(\widehat{F}_n), T_2(\widehat{F}_n), T_3(\widehat{F}_n), T_4(\widehat{F}_n), T_5(\widehat{F}_n)).$$

We get

$$\widehat{\rho} = \frac{\sum_{i=1}^{n} (X_i - \overline{X}_n)(Y_i - \overline{Y}_n)}{\sqrt{\sum_{i=1}^{n} (X_i - \overline{X}_n)^2} \sqrt{\sum_{i=1}^{n} (Y_i - \overline{Y}_n)^2}}$$

which is called the **sample correlation**. ∎

2.17 Example (Quantiles). Let F be strictly increasing with density f. Let $T(F) = F^{-1}(p)$ be the pth quantile. The estimate of $T(F)$ is $\widehat{F}_n^{-1}(p)$. We have to be a bit careful since $\widehat{F}_n$ is not invertible. To avoid ambiguity we define $\widehat{F}_n^{-1}(p) = \inf\{x : \widehat{F}_n(x) \geq p\}$. We call $\widehat{F}_n^{-1}(p)$ the pth **sample quantile**. ∎

The Glivenko–Cantelli theorem ensures that $\widehat{F}_n$ converges to F. This suggests that $\widehat{\theta}_n = T(\widehat{F}_n)$ will converge to $\theta = T(F)$. Furthermore, we would hope that under reasonable conditions, $\widehat{\theta}_n$ will be asymptotically normal. This leads us to the next topic.

2.3   Influence Functions

The influence function is used to approximate the standard error of a plug-in estimator. The formal definition is as follows.

2.18 Definition. *The* **Gâteaux derivative** *of T at F in the direction G is defined by*

$$L_F(G) = \lim_{\epsilon \to 0} \frac{T\big((1-\epsilon)F + \epsilon G\big) - T(F)}{\epsilon}. \tag{2.19}$$

If $G = \delta_x$ is a point mass at x then we write $L_F(x) \equiv L_F(\delta_x)$ and we call $L_F(x)$ the **influence function**. *Thus,*

$$L_F(x) = \lim_{\epsilon \to 0} \frac{T\big((1-\epsilon)F + \epsilon\delta_x\big) - T(F)}{\epsilon}. \tag{2.20}$$

The **empirical influence function** *is defined by $\widehat{L}(x) = L_{\widehat{F}_n}(x)$. Thus,*

$$\widehat{L}(x) = \lim_{\epsilon \to 0} \frac{T\big((1-\epsilon)\widehat{F}_n + \epsilon\delta_x\big) - T(\widehat{F}_n)}{\epsilon}. \tag{2.21}$$

Often we drop the subscript F and write $L(x)$ instead of $L_F(x)$.

2.22 Theorem. *Let $T(F) = \int a(x)dF(x)$ be a linear functional. Then:*

1. $L_F(x) = a(x) - T(F)$ *and* $\widehat{L}(x) = a(x) - T(\widehat{F}_n)$.

2. *For any G,*

$$T(G) = T(F) + \int L_F(x)dG(x). \tag{2.23}$$

3. $\int L_F(x)dF(x) = 0$.

4. *Let* $\tau^2 = \int L_F^2(x)dF(x)$. *Then,* $\tau^2 = \int(a(x) - T(F))^2 dF(x)$ *and if* $\tau^2 < \infty$,

$$\sqrt{n}(T(F) - T(\widehat{F}_n)) \rightsquigarrow N(0, \tau^2). \tag{2.24}$$

5. *Let*

$$\widehat{\tau}^2 = \frac{1}{n}\sum_{i=1}^{n}\widehat{L}^2(X_i) = \frac{1}{n}\sum_{i=1}^{n}(a(X_i) - T(\widehat{F}_n))^2. \tag{2.25}$$

Then, $\widehat{\tau}^2 \xrightarrow{P} \tau^2$ *and* $\widehat{se}/se \xrightarrow{P} 1$ *where* $\widehat{se} = \widehat{\tau}/\sqrt{n}$ *and* $se = \sqrt{\mathbb{V}(T(\widehat{F}_n))}$.

6. We have that

$$\frac{\sqrt{n}(T(F) - T(\widehat{F}_n))}{\widehat{\tau}} \rightsquigarrow N(0, 1). \tag{2.26}$$

PROOF. The first three claims follow easily from the definition of the influence function. To prove the fourth claim, write

$$
\begin{aligned}
T(\widehat{F}_n) &= T(F) + \int L_F(x) d\widehat{F}_n(x) \\
&= T(F) + \frac{1}{n} \sum_{i=1}^n L_F(X_i).
\end{aligned}
$$

From the central limit theorem and the fact that $\int L_F(x) dF(x) = 0$, it follows that

$$\sqrt{n}(T(F) - T(\widehat{F}_n)) \rightsquigarrow N(0, \tau^2)$$

where $\tau^2 = \int L_F^2(x) dF(x)$. The fifth claim follows from the law of large numbers. The final statement follows from the fourth and fifth claims and Slutsky's theorem. ■

The theorem above tells us that the influence function $L_F(x)$ behaves like the score function in parametric estimation. To see this, recall that if $f(x; \theta)$ is a parametric model, $\mathcal{L}_n(\theta) = \prod_{i=1}^n f(X_i; \theta)$ is the likelihood function and the maximum likelihood estimator $\widehat{\theta}_n$ is the value of θ that maximizes $\mathcal{L}_n(\theta)$. The score function is $s_\theta(x) = \partial \log f(x; \theta)/\partial\theta$ which, under appropriate regularity conditions, satisfies $\int s_\theta(x) f(x; \theta) dx = 0$ and $\mathbb{V}(\widehat{\theta}_n) \approx \int (s_\theta(x))^2 f(x; \theta) dx/n$. Similarly, for the influence function we have that $\int L_F(x) dF(x) = 0$ and and $\mathbb{V}(T(\widehat{F}_n)) \approx \int L_F^2(x) dF(x)/n$.

If the functional $T(F)$ is not linear, then (2.23) will not hold exactly, but it may hold approximately.

2.27 Theorem. *If T is Hadamard differentiable[2] with respect to $d(F, G) = \sup_x |F(x) - G(x)|$ then*

$$\sqrt{n}(T(\widehat{F}_n) - T(F)) \rightsquigarrow N(0, \tau^2) \tag{2.28}$$

where $\tau^2 = \int L_F(x)^2 dF(x)$. Also,

$$\frac{(T(\widehat{F}_n) - T(F))}{\widehat{se}} \rightsquigarrow N(0, 1) \tag{2.29}$$

---

[2]Hadamard differentiability is defined in the appendix.

where $\widehat{se} = \widehat{\tau}/\sqrt{n}$ and

$$\widehat{\tau} = \frac{1}{n} \sum_{i=1}^{n} L^2(X_i). \tag{2.30}$$

We call the approximation $(T(\widehat{F}_n) - T(F))/\widehat{se} \approx N(0,1)$ the **nonparametric delta method**. From the normal approximation, a large sample confidence interval is $T(\widehat{F}_n) \pm z_{\alpha/2}\,\widehat{se}$. This is only a pointwise asymptotic confidence interval. In summary:

The Nonparametric Delta Method

A $1 - \alpha$, pointwise asymptotic confidence interval for $T(F)$ is

$$T(\widehat{F}_n) \pm z_{\alpha/2}\,\widehat{se} \tag{2.31}$$

where

$$\widehat{se} = \frac{\widehat{\tau}}{\sqrt{n}} \quad \text{and} \quad \widehat{\tau}^2 = \frac{1}{n}\sum_{i=1}^{n} \widehat{L}^2(X_i).$$

2.32 Example (The mean). Let $\theta = T(F) = \int x\, dF(x)$. The plug-in estimator is $\widehat{\theta} = \int x\, d\widehat{F}_n(x) = \overline{X}_n$. Also, $T((1 - \epsilon)F + \epsilon\delta_x) = (1 - \epsilon)\theta + \epsilon x$. Thus, $L(x) = x - \theta$, $\widehat{L}(x) = x - \overline{X}_n$ and $\widehat{se}^2 = \widehat{\sigma}^2/n$ where $\widehat{\sigma}^2 = n^{-1}\sum_{i=1}^{n}(X_i - \overline{X}_n)^2$. A pointwise asymptotic nonparametric 95 percent confidence interval for θ is $\overline{X}_n \pm 2\,\widehat{se}$. ∎

Sometimes statistical functionals take the form $T(F) = a(T_1(F), \ldots, T_m(F))$ for some function $a(t_1, \ldots, t_m)$. By the chain rule, the influence function is

$$L(x) = \sum_{i=1}^{m} \frac{\partial a}{\partial t_i} L_i(x)$$

where

$$L_i(x) = \lim_{\epsilon \to 0} \frac{T_i((1 - \epsilon)F + \epsilon\delta_x) - T_i(F)}{\epsilon}. \tag{2.33}$$

2.34 Example (Correlation). Let $Z = (X, Y)$ and let $T(F) = E(X - \mu_X)(Y - \mu_Y)/(\sigma_x\sigma_y)$ denote the correlation where $F(x,y)$ is bivariate. Recall that $T(F) = a(T_1(F), T_2(F), T_3(F), T_4(F), T_5(F))$ where

$$T_1(F) = \int x\, dF(z) \quad T_2(F) = \int y\, dF(z) \quad T_3(F) = \int xy\, dF(z)$$
$$T_4(F) = \int x^2\, dF(z) \quad T_5(F) = \int y^2\, dF(z)$$

and

$$a(t_1, \ldots, t_5) = \frac{t_3 - t_1 t_2}{\sqrt{(t_4 - t_1^2)(t_5 - t_2^2)}}.$$

It follows from (2.33) that

$$L(x, y) = \widetilde{x}\widetilde{y} - \frac{1}{2}T(F)(\widetilde{x}^2 + \widetilde{y}^2)$$

where

$$\widetilde{x} = \frac{x - \int x dF}{\sqrt{\int x^2 dF - (\int x dF)^2}}, \quad \widetilde{y} = \frac{y - \int y dF}{\sqrt{\int y^2 dF - (\int y dF)^2}}. \quad \blacksquare$$

2.35 Example (Quantiles). Let F be strictly increasing with positive density f. The $T(F) = F^{-1}(p)$ be the pth quantile. The influence function is (see Exercise 10)

$$L(x) = \begin{cases} \frac{p-1}{f(\theta)}, & x \leq \theta \\ \frac{p}{f(\theta)}, & x > \theta. \end{cases}$$

The asymptotic variance of $T(\widehat{F}_n)$ is

$$\frac{\tau^2}{n} = \frac{1}{n} \int L^2(x) dF(x) = \frac{p(1-p)}{nf^2(\theta)}. \tag{2.36}$$

To estimate this variance we need to estimate the density f. Later we shall see that the bootstrap provides a simpler estimate of the variance. $\blacksquare$

2.4   Empirical Probability Distributions

This section discusses a generalization of the DKW inequality. The reader may skip this section if desired. Using the empirical CDF to estimate the true CDF is a special case of a more general idea. Let $X_1, \ldots, X_n \sim P$ be an IID sample from a probability measure P. Define the **empirical probability distribution** $\mathbb{P}_n$ by

$$\widehat{P}_n(A) = \frac{\text{number of } X_i \in A}{n}. \tag{2.37}$$

We would like to be able to say that $\mathbb{P}_n$ is close to P in some sense. For a fixed A we know that $n\widehat{P}_n(A) \sim \text{Binomial}(n, p)$ where $p = P(A)$. By Hoeffding's inequality, it follows that

$$\mathbb{P}(|\widehat{P}_n(A) - P(A)| > \epsilon) \leq 2e^{-2n\epsilon^2}. \tag{2.38}$$

We would like to extend this to be a statement of the form

$$\mathbb{P}\left(\sup_{A \in \mathcal{A}} |\widehat{P}_n(A) - P(A)| > \epsilon\right) \leq \text{something small}$$

for some class of sets $\mathcal{A}$. This is exactly what the DKW inequality does by taking $\mathcal{A} = \{A = (-\infty, t] : t \in \mathbb{R}\}$. But DKW is only useful for one-dimensional random variables. We can get a more general inequality by using Vapnik–Chervonenkis (VC) theory.

Let $\mathcal{A}$ be a class of sets. Given a finite set $R = \{x_1, \ldots, x_n\}$ let

$$N_{\mathcal{A}}(R) = \#\left\{ R \bigcap A : A \in \mathcal{A} \right\} \tag{2.39}$$

be the number of subsets of R "picked out" as A varies over $\mathcal{A}$. We say that R is **shattered** by $\mathcal{A}$ if $N_{\mathcal{A}}(R) = 2^n$. The **shatter coefficient** is defined by

$$s(\mathcal{A}, n) = \max_{R \in \mathcal{F}_n} N_{\mathcal{A}}(R) \tag{2.40}$$

where $\mathcal{F}_n$ consists of all finite sets of size n.

2.41 Theorem (Vapnik and Chervonenkis, 1971). *For any P, n and $\epsilon > 0$,*

$$\mathbb{P}\left(\sup_{A \in \mathcal{A}} |\widehat{P}_n(A) - P(A)| > \epsilon \right) \leq 8s(\mathcal{A}, n)e^{-n\epsilon^2/32}. \tag{2.42}$$

Theorem 2.41 is only useful if the shatter coefficients do not grow too quickly with n. This is where VC dimension enters. If $s(\mathcal{A}, n) = 2^n$ for all n set $\mathrm{VC}(\mathcal{A}) = \infty$. Otherwise, define $\mathrm{VC}(\mathcal{A})$ to be the largest k for which $s(\mathcal{A}, k) = 2^k$. We call $\mathrm{VC}(\mathcal{A})$ the **Vapnik–Chervonenkis** dimension of $\mathcal{A}$. Thus, the VC-dimension is the size of the largest finite set F that is shattered by $\mathcal{A}$. The following theorem shows that if $\mathcal{A}$ has finite VC-dimension then the shatter coefficients grow as a polynomial in n.

2.43 Theorem. *If $\mathcal{A}$ has finite VC-dimension v, then*

$$s(\mathcal{A}, n) \leq n^v + 1.$$

In this case,

$$\mathbb{P}\left(\sup_{A \in \mathcal{A}} |\widehat{P}_n(A) - P(A)| > \epsilon \right) \leq 8(n^v + 1)e^{-n\epsilon^2/32}. \tag{2.44}$$

2.45 Example. Let $\mathcal{A} = \{(-\infty, x]; x \in \mathbb{R}\}$. Then $\mathcal{A}$ shatters every one point set $\{x\}$ but it shatters no set of the form $\{x, y\}$. Therefore, $\mathrm{VC}(\mathcal{A}) = 1$. Since, $\mathbb{P}((-\infty, x]) = F(x)$ is the CDF and $\widehat{P}_n((-\infty, x]) = \widehat{F}_n(x)$ is the empirical CDF, we conclude that

$$\mathbb{P}\left(\sup_x |\widehat{F}_n(x) - F(x)| > \epsilon \right) \leq 8(n + 1)e^{-n\epsilon^2/32}$$

which is looser than the DKW bound. This shows that the bound (2.42) is not the tightest possible. ∎

2.46 Example. Let $\mathcal{A}$ be the set of closed intervals on the real line. Then $\mathcal{A}$ shatters $S = \{x, y\}$ but it cannot shatter sets with three points. Consider $S = \{x, y, z\}$ where $x < y < z$. One cannot find an interval A such that $A \cap S = \{x, z\}$. So, $VC(\mathcal{A}) = 2$. $\blacksquare$

2.47 Example. Let $\mathcal{A}$ be all linear half-spaces on the plane. Any three-point set (not all on a line) can be shattered. No four-point set can be shattered. Consider, for example, four points forming a diamond. Let T be the leftmost and rightmost points. This set cannot be picked out. Other configurations can also be seen to be unshatterable. So $VC(\mathcal{A}) = 3$. In general, halfspaces in $\mathcal{R}^d$ have VC dimension $d + 1$. $\blacksquare$

2.48 Example. Let $\mathcal{A}$ be all rectangles on the plane with sides parallel to the axes. Any four-point set can be shattered. Let S be a five-point set. There is one point that is not leftmost, rightmost, uppermost or lowermost. Let T be all points in S except this point. Then T can't be picked out. So, we have that $VC(\mathcal{A}) = 4$. $\blacksquare$

2.5   Bibliographic Remarks

Further details on statistical functionals can be found in Serfling (1980), Davison and Hinkley (1997), Shao and Tu (1995), Fernholz (1983) and van der Vaart (1998). Vapnik–Chervonenkis theory is discussed in Devroye et al. (1996), van der Vaart (1998) and van der Vaart and Wellner (1996).

2.6   Appendix

Here are some details about Theorem 2.27. Let $\mathfrak{F}$ denote all distribution functions and let $\mathcal{D}$ denote the linear space generated by $\mathfrak{F}$. Write $T((1 - \epsilon)F + \epsilon G) = T(F + \epsilon D)$ where $D = G - F \in \mathcal{D}$. The Gateâux derivative, which we now write as $L_F(D)$, is defined by

$$\lim_{\epsilon \downarrow 0} \left| \frac{T(F + \epsilon D) - T(F)}{\epsilon} - L_F(D) \right| \to 0.$$

Thus $T(F + \epsilon D) \approx \epsilon L_F(D) + o(\epsilon)$ and the error term $o(\epsilon)$ goes to 0 as $\epsilon \to 0$. Hadamard differentiability requires that this error term be small uniformly over compact sets. Equip $\mathcal{D}$ with a metric d. T is **Hadamard differentiable** at F if there exists a linear functional L_F on $\mathcal{D}$ such that for any $\epsilon_n \to 0$ and $\{D, D_1, D_2, \ldots\} \subset \mathcal{D}$ such that $d(D_n, D) \to 0$ and $F + \epsilon_n D_n \in \mathcal{F}$,

$$\lim_{n \to \infty} \left(\frac{T(F + \epsilon_n D_n) - T(F)}{\epsilon_n} - L_F(D_n) \right) = 0.$$

2.7   Exercises

1. Fill in the details of the proof of Theorem 2.22.

2. Prove Theorem 2.4.

3. (Computer experiment.) Generate 100 observations from a N(0,1) distribution. Compute a 95 percent confidence band for the CDF F. Repeat this 1000 times and see how often the confidence band contains the true distribution function. Repeat using data from a Cauchy distribution.

4. Let $X_1, \ldots, X_n \sim F$ and let $\widehat{F}_n(x)$ be the empirical distribution function. For a fixed x, find the limiting distribution of $\sqrt{\widehat{F}_n(x)}$.

5. Suppose that
$$|T(F) - T(G)| \le C \, ||F - G||_\infty \tag{2.49}$$
for some constant $0 < C < \infty$ where $||F - G||_\infty = \sup_x |F(x) - G(x)|$. Prove that $T(\widehat{F}_n) \xrightarrow{\text{a.s.}} T(F)$. Suppose that $|X| \le M < \infty$. Show that $T(F) = \int x \, dF(x)$ satisfies (2.49).

6. Let x and y be two distinct points. Find $\text{Cov}(\widehat{F}_n(x), \widehat{F}_n(y))$.

7. Let $X_1, \ldots, X_n \sim \text{Bernoulli}(p)$ and let $Y_1, \ldots, Y_m \sim \text{Bernoulli}(q)$. Find the plug-in estimator and estimated standard error for p. Find an approximate 90 percent confidence interval for p. Find the plug-in estimator and estimated standard error for $p - q$. Find an approximate 90 percent confidence interval for $p - q$.

8. Let $X_1, \ldots, X_n \sim F$ and let $\widehat{F}$ be the empirical distribution function. Let $a < b$ be fixed numbers and define $\theta = T(F) = F(b) - F(a)$. Let $\widehat{\theta} = T(\widehat{F}_n) = \widehat{F}_n(b) - \widehat{F}_n(a)$. Find the influence function. Find the estimated standard error of $\widehat{\theta}$. Find an expression for an approximate $1 - \alpha$ confidence interval for θ.

9. Verify the formula for the influence function in Example 2.34.

10. Verify the formula for the influence function in Example 2.35. *Hint*: Let $F_\epsilon(y) = (1 - \epsilon)F(y) + \epsilon\delta_x(y)$ where $\delta_x(y)$ is a point mass at x, i.e., $\delta_x(y) = 0$ if $y < x$ and $\delta_x(y) = 1$ if $y \geq x$. By the definition of $T(F)$, we have that $p = F_\epsilon(T(F_\epsilon))$. Now differentiate with respect to ϵ and evaluate the derivative at $\epsilon = 0$.

11. Data on the magnitudes of earthquakes near Fiji are available on the book website. Estimate the CDF $F(x)$. Compute and plot a 95 percent confidence envelope for F. Find an approximate 95 percent confidence interval for $F(4.9) - F(4.3)$.

12. Get the data on eruption times and waiting times between eruptions of the Old Faithful geyser from the book website. Estimate the mean waiting time and give a standard error for the estimate. Also, give a 90 percent confidence interval for the mean waiting time. Now estimate the median waiting time. In the next chapter we will see how to get the standard error for the median.

13. In 1975, an experiment was conducted to see if cloud seeding produced rainfall. 26 clouds were seeded with silver nitrate and 26 were not. The decision to seed or not was made at random. Get the data from

    http://lib.stat.cmu.edu/DASL/Stories/CloudSeeding.html

    Let $\theta = T(F_1) - T(F_2)$ be the difference in the median precipitation from the two groups. Estimate θ. Estimate the standard error of the estimate and produce a 95 percent confidence interval. To estimate the standard error you will need to use formula (2.36). This formula requires the density f so you will need to insert an estimate of f. What will you do? Be creative.

14. Let $\mathcal{A}$ be the set of two-dimensional spheres. That is, $A \in \mathcal{A}$ if $A = \{(x, y) : (x-a)^2 + (y-b)^2 \leq c^2\}$ for some a, b, c. Find the VC dimension of $\mathcal{A}$.

15. The empirical CDF can be regarded as a nonparametric maximum likelihood estimator. For example, consider data $X_1, \ldots, X_n$ on [0,1]. Divide the interval into bins of width Δ and find the MLE over all distributions with constant density over the bins. Show that the resulting CDF converges to the empirical CDF as $\Delta \to 0$.

3
The Bootstrap and the Jackknife

The bootstrap and the jackknife are nonparametric methods for computing standard errors and confidence intervals. The jackknife is less computationally expensive, but the bootstrap has some statistical advantages.

3.1 The Jackknife

The jackknife, due to Quenouille (1949), is a simple method for approximating the bias and variance of an estimator. Let $T_n = T(X_1, \ldots, X_n)$ be an estimator of some quantity θ and let $\mathsf{bias}(T_n) = \mathbb{E}(T_n) - \theta$ denote the bias. Let $T_{(-i)}$ denote the statistic with the ith observation removed. The **jackknife bias estimate** is defined by

$$b_{\text{jack}} = (n-1)(\overline{T}_n - T_n) \tag{3.1}$$

where $\overline{T}_n = n^{-1} \sum_i T_{(-i)}$. The bias-corrected estimator is $T_{\text{jack}} = T_n - b_{\text{jack}}$. Why is b_{jack} defined this way? For many statistics it can be shown that

$$\mathsf{bias}(T_n) = \frac{a}{n} + \frac{b}{n^2} + O\left(\frac{1}{n^3}\right) \tag{3.2}$$

for some a and b. For example, let $\sigma^2 = \mathbb{V}(X_i)$ and let $\widehat{\sigma}_n^2 = n^{-1} \sum_{i=1}^n (X_i - \overline{X})^2$. Then, $\mathbb{E}(\widehat{\sigma}_n^2) = (n-1)\sigma^2/n$ so that $\mathsf{bias}(\widehat{\sigma}_n^2) = -\sigma^2/n$. Thus, (3.2) holds with $a = -\sigma^2$ and $b = 0$.

When (3.2) holds, we have

$$\text{bias}(T_{(-i)}) = \frac{a}{n-1} + \frac{b}{(n-1)^2} + O\left(\frac{1}{n^3}\right).\tag{3.3}$$

It follows that $\text{bias}(\overline{T}_n)$ also satisfies (3.3). Hence,

$$
\begin{aligned}
\mathbb{E}(b_{\text{jack}}) &= (n-1)\big(\mathbb{E}(\text{bias}(\overline{T}_n)) - \mathbb{E}(\text{bias}(T_n))\big)\\
&= (n-1)\left[\left(\frac{1}{n-1} - \frac{1}{n}\right)a + \left(\frac{1}{(n-1)^2} - \frac{1}{n^2}\right)b + O\left(\frac{1}{n^3}\right)\right]\\
&= \frac{a}{n} + \frac{(2n-1)b}{n^2(n-1)} + O\left(\frac{1}{n^2}\right)\\
&= \text{bias}(T_n) + O\left(\frac{1}{n^2}\right)
\end{aligned}
$$

which shows that b_{jack} estimates the bias up to order $O(n^{-2})$. By a similar calculation,

$$\text{bias}(T_{\text{jack}}) = -\frac{b}{n(n-1)} + O\left(\frac{1}{n^2}\right) = O\left(\frac{1}{n^2}\right)$$

so the bias of T_{jack} is an order of magnitude smaller than that of T_n. T_{jack} can also be written as

$$T_{\text{jack}} = \frac{1}{n}\sum_{i=1}^n \widetilde{T}_i$$

where

$$\widetilde{T}_i = nT_n - (n-1)T_{(-i)}$$

are called **pseudo-values**.

The jackknife estimate of $\mathbb{V}(T_n)$ is

$$v_{\text{jack}} = \frac{\widetilde{s}^2}{n}\tag{3.4}$$

where

$$\widetilde{s}^2 = \frac{\sum_{i=1}^n \left(\widetilde{T}_i - \frac{1}{n}\sum_{i=1}^n \widetilde{T}_i\right)^2}{n-1}$$

is the sample variance of the pseudo-values. Under suitable conditions on T, it can be shown that v_{jack} consistently estimates $\mathbb{V}(T_n)$. For example, if T is a smooth function of the sample mean, then consistency holds.

3.5 Theorem. *Let $\mu = \mathbb{E}(X_1)$ and $\sigma^2 = \mathbb{V}(X_1) < \infty$ and suppose that $T_n = g(\overline{X}_n)$ where g has a continuous, nonzero derivative at μ. Then $(T_n - $*

$g(\mu))/\sigma_n \rightsquigarrow N(0,1)$ *where* $\sigma_n^2 = n^{-1}(g'(\mu))^2\sigma^2$. *The jackknife is consistent, meaning that*

$$\frac{v_{\text{jack}}}{\sigma_n^2} \xrightarrow{\text{a.s.}} 1. \tag{3.6}$$

3.7 Theorem (Efron, 1982). *If* $T(F) = F^{-1}(p)$ *is the* p^{th} *quantile, then the jackknife variance estimate is inconsistent. For the median* $(p = 1/2)$ *we have that* $v_{\text{jack}}/\sigma_n^2 \rightsquigarrow (\chi_2^2/2)^2$ *where* σ_n^2 *is the asymptotic variance of the sample median.*

3.8 Example. Let $T_n = \overline{X}_n$. It is easy to see that $\widetilde{T}_i = X_i$. Hence, $T_{\text{jack}} = T_n$, $b = 0$ and $v_{\text{jack}} = S_n^2/n$ where S_n^2 is the sample variance. ∎

There is a connection between the jackknife and the influence function. Recall that the influence function is

$$L_F(x) = \lim_{\epsilon \to 0} \frac{T((1-\epsilon)F + \epsilon\delta_x) - T(F)}{\epsilon}. \tag{3.9}$$

Suppose we approximate $L_F(X_i)$ by setting $F = \widehat{F}_n$ and $\epsilon = -1/(n-1)$. This yields the approximation

$$L_F(X_i) \approx \frac{T((1-\epsilon)\widehat{F}_n + \epsilon\delta_{x_i}) - T(\widehat{F}_n)}{\epsilon} = (n-1)(T_n - T_{(-i)}) \equiv \ell_i.$$

It follows that

$$b = -\frac{1}{n}\sum_{i=1}^{n}\ell_i, \quad v_{\text{jack}} = \frac{1}{n(n-1)}\left(\sum_i \ell_i^2 - nb^2\right).$$

In other words, the jackknife is an approximate version of the nonparametric delta method.

3.10 Example. Consider estimating the skewness $T(F) = \int(x-\mu)^3 dF(x)/\sigma^3$ of the nerve data. The point estimate is $T(\widehat{F}_n) = 1.76$. The jackknife estimate of the standard error is .17. An approximate 95 percent confidence interval for $T(F)$ is $1.76 \pm 2(.17) = (1.42, 2.10)$. These exclude 0 which suggests that the data are not Normal. We can also compute the standard error using the influence function. For this functional, we have (see Exercise 1)

$$L_F(x) = \frac{(x-\mu)^3}{\sigma^3} - T(F)\left(1 + \frac{3}{2}\frac{((x-\mu)^2 - \sigma^2)}{\sigma^2}\right).$$

Then

$$\widehat{\text{se}} = \sqrt{\frac{\widehat{\tau}^2}{n}} = \sqrt{\frac{\sum_{i=1}^{n}\widehat{L}^2(X_i)}{n^2}} = .18.$$

It is reassuring to get nearly the same answer. ∎

3.2   The Bootstrap

The bootstrap is a method for estimating the variance and the distribution of a statistic $T_n = g(X_1, \ldots, X_n)$. We can also use the bootstrap to construct confidence intervals.

Let $\mathbb{V}_F(T_n)$ denote the variance of T_n. We have added the subscript F to emphasize that the variance is a function of F. If we knew F we could, at least in principle, compute the variance. For example, if $T_n = n^{-1} \sum_{i=1}^n X_i$, then

$$\mathbb{V}_F(T_n) = \frac{\sigma^2}{n} = \frac{\int x^2 dF(x) - \left(\int x dF(x) \right)^2}{n}$$

which is clearly a function of F.

With the bootstrap, we estimate $\mathbb{V}_F(T_n)$ with $\mathbb{V}_{\widehat{F}_n}(T_n)$. In other words, we use a plug-in estimator of the variance. Since, $\mathbb{V}_{\widehat{F}_n}(T_n)$ may be difficult to compute, we approximate it with a simulation estimate denoted by v_{boot}. Specifically, we do the following steps:

Bootstrap Variance Estimation

1. Draw $X_1^*, \ldots, X_n^* \sim \widehat{F}_n$.

2. Compute $T_n^* = g(X_1^*, \ldots, X_n^*)$.

3. Repeat steps 1 and 2, B times to get $T_{n,1}^*, \ldots, T_{n,B}^*$.

4. Let

$$v_{\text{boot}} = \frac{1}{B} \sum_{b=1}^B \left(T_{n,b}^* - \frac{1}{B} \sum_{r=1}^B T_{n,r}^* \right)^2. \tag{3.11}$$

By the law of large numbers, $v_{\text{boot}} \xrightarrow{\text{a.s.}} \mathbb{V}_{\widehat{F}_n}(T_n)$ as $B \to \infty$. The estimated standard error of T_n is $\widehat{\text{se}}_{\text{boot}} = \sqrt{v_{\text{boot}}}$. The following diagram illustrates the bootstrap idea:

$$
\begin{array}{llll}
\text{Real world:} & F & \Longrightarrow & X_1, \ldots, X_n & \Longrightarrow & T_n = g(X_1, \ldots, X_n) \\
\text{Bootstrap world:} & \widehat{F}_n & \Longrightarrow & X_1^*, \ldots, X_n^* & \Longrightarrow & T_n^* = g(X_1^*, \ldots, X_n^*)
\end{array}
$$

$$\mathbb{V}_F(T_n) \overset{O(1/\sqrt{n})}{\approx} \mathbb{V}_{\widehat{F}_n}(T_n) \overset{O(1/\sqrt{B})}{\approx} v_{\text{boot}}.$$

How do we simulate from $\widehat{F}_n$? Since $\widehat{F}_n$ gives probability $1/n$ to each data point, **drawing n points at random from $\widehat{F}_n$ is the same as drawing a**

```
                    Bootstrap for the Median
Given data X = (X(1), ..., X(n)):

T     = median(X)
Tboot = vector of length B
for(i in 1:N){
    Xstar    = sample of size n from X (with replacement)
    Tboot[i] = median(Xstar)
    }
se = sqrt(variance(Tboot))
```

FIGURE 3.1. Pseudo-code for bootstrapping the median.

sample of size n with replacement from the original data. Therefore step 1 can be replaced by:

1. Draw $X_1^*, \ldots, X_n^*$ with replacement from $X_1, \ldots, X_n$.

3.12 Example. Figure 3.1 shows pseudo-code for using the bootstrap to estimate the standard error of the median. ∎

The bootstrap can be used to approximate the CDF of a statistic T_n. Let $G_n(t) = \mathbb{P}(T_n \leq t)$ be the CDF of T_n. The bootstrap approximation to G_n is

$$\widehat{G}_n^*(t) = \frac{1}{B} \sum_{b=1}^{B} I\left(T_{n,b}^* \leq t\right). \tag{3.13}$$

3.3 Parametric Bootstrap

So far, we have estimated F nonparametrically. There is also a **parametric bootstrap**. If F_θ depends on a parameter θ and $\widehat{\theta}$ is an estimate of θ, then we simply sample from $F_{\widehat{\theta}}$ instead of $\widehat{F}_n$. This is just as accurate, but much simpler than, the delta method.

3.14 Example. When applied to the nerve data, the bootstrap, based on $B = 1000$ replications, yields a standard error for the estimated skewness of .16 which is nearly identical to the jackknife estimate. ∎

3.4   Bootstrap Confidence Intervals

There are several ways to construct bootstrap confidence intervals. They vary in ease of calculation and accuracy.

Normal Interval. The simplest is the Normal interval

$$T_n \pm z_{\alpha/2}\, \widehat{\text{se}}_{\text{boot}}$$

where $\widehat{\text{se}}_{\text{boot}}$ is the bootstrap estimate of the standard error. This interval is not accurate unless the distribution of T_n is close to Normal.

Pivotal Intervals.   Let $\theta = T(F)$ and $\widehat{\theta}_n = T(\widehat{F}_n)$ and define the **pivot** $R_n = \widehat{\theta}_n - \theta$. Let $H(r)$ denote the CDF of the pivot:

$$H(r) = \mathbb{P}_F(R_n \leq r).$$

Let $C_n^\star = (a, b)$ where

$$a = \widehat{\theta}_n - H^{-1}\left(1 - \frac{\alpha}{2}\right) \quad \text{and} \quad b = \widehat{\theta}_n - H^{-1}\left(\frac{\alpha}{2}\right).$$

It follows that

$$
\begin{aligned}
\mathbb{P}(a \leq \theta \leq b) &= \mathbb{P}(\widehat{\theta}_n - b \leq R_n \leq \widehat{\theta}_n - a) \\
&= H(\widehat{\theta}_n - a) - H(\widehat{\theta}_n - b) \\
&= H\left(H^{-1}\left(1 - \frac{\alpha}{2}\right)\right) - H\left(H^{-1}\left(\frac{\alpha}{2}\right)\right) \\
&= 1 - \frac{\alpha}{2} - \frac{\alpha}{2} = 1 - \alpha.
\end{aligned}
$$

Hence, $C_n^\star$ is an exact $1 - \alpha$ confidence interval for θ. Unfortunately, a and b depend on the unknown distribution H but we can form a bootstrap estimate of H:

$$\widehat{H}(r) = \frac{1}{B}\sum_{b=1}^{B} I(R_{n,b}^* \leq r)$$

where $R_{n,b}^* = \widehat{\theta}_{n,b}^* - \widehat{\theta}_n$. Let r_β^* denote the β sample quantile of $(R_{n,1}^*, \ldots, R_{n,B}^*)$ and let θ_β^* denote the β sample quantile of $(\theta_{n,1}^*, \ldots, \theta_{n,B}^*)$. Note that $r_\beta^* = \theta_\beta^* - \widehat{\theta}_n$. It follows that an approximate $1 - \alpha$ confidence interval is $C_n = (\widehat{a}, \widehat{b})$ where

$$
\begin{aligned}
\widehat{a} &= \widehat{\theta}_n - \widehat{H}^{-1}\left(1 - \frac{\alpha}{2}\right) = \widehat{\theta}_n - r_{1-\alpha/2}^* = 2\widehat{\theta}_n - \theta_{1-\alpha/2}^* \\
\widehat{b} &= \widehat{\theta}_n - \widehat{H}^{-1}\left(\frac{\alpha}{2}\right) \quad\quad = \widehat{\theta}_n - r_{\alpha/2}^* \quad = 2\widehat{\theta}_n - \theta_{\alpha/2}^*.
\end{aligned}
$$

In summary:

The $1 - \alpha$ **bootstrap pivotal confidence** interval is

$$C_n = \left(2\widehat{\theta}_n - \widehat{\theta}^*_{((1-\alpha/2)B)}, \ 2\widehat{\theta}_n - \widehat{\theta}^*_{((\alpha/2)B)} \right). \tag{3.15}$$

Typically, this is a pointwise, asymptotic confidence interval.

The next result follows from Theorem 3.21.

3.16 Theorem. *If $T(F)$ is Hadamard differentiable and C_n is given in (3.15) then $\mathbb{P}_F(T(F) \in C_n) \to 1 - \alpha$.*

Studentized Pivotal Interval. There is a different version of the pivotal interval which has some advantages. Let

$$Z_n = \frac{T_n - \theta}{\widehat{se}_{boot}}$$

and

$$Z^*_{n,b} = \frac{T^*_{n,b} - T_n}{\widehat{se}^*_b}$$

where $\widehat{se}^*_b$ is an estimate of the standard error of $T^*_{n,b}$ **not** T_n. Now we reason as in the pivotal interval. The sample quantiles of the bootstrap quantities $Z^*_{n,1}, \ldots, Z^*_{n,B}$ should approximate the true quantiles of the distribution of Z_n. Let z^*_α denote the α sample quantile of $Z^*_{n,1}, \ldots, Z^*_{n,B}$, then $\mathbb{P}(Z_n \le z^*_\alpha) \approx \alpha$. Let

$$C_n = \left(T_n - z^*_{1-\alpha/2}\widehat{se}_{boot}, \ T_n - z^*_{\alpha/2}\widehat{se}_{boot} \right).$$

Then,

$$
\begin{aligned}
\mathbb{P}(\theta \in C_n) &= \mathbb{P}\left(T_n - z^*_{1-\alpha/2}\widehat{se}_{boot} \le \theta \le T_n - z^*_{\alpha/2}\widehat{se}_{boot} \right) \\
&= \mathbb{P}\left(z^*_{\alpha/2} \le \frac{T_n - \theta}{se_{boot}} \le z^*_{1-\alpha/2} \right) \\
&= \mathbb{P}\left(z^*_{\alpha/2} \le Z_n \le z^*_{1-\alpha/2} \right) \\
&\approx 1 - \alpha.
\end{aligned}
$$

This interval has higher accuracy then all the intervals discussed so far (see Section 3.5) but there is a catch: you need to compute $\widehat{se}^*_b$ for each bootstrap sample. This may require doing a second bootstrap within each bootstrap.

The $1 - \alpha$ **bootstrap studentized pivotal interval** is

$$\left(T_n - z^*_{1-\alpha/2}\, \widehat{\mathrm{se}}_{\mathrm{boot}},\ T_n - z^*_{\alpha/2}\, \widehat{\mathrm{se}}_{\mathrm{boot}} \right)$$

where z^*_β is the β quantile of $Z^*_{n,1}, \ldots, Z^*_{n,B}$ and

$$Z^*_{n,b} = \frac{T^*_{n,b} - T_n}{\widehat{\mathrm{se}}^*_b}.$$

Percentile Intervals. The **bootstrap percentile interval** is defined by

$$C_n = \left(T^*_{(B\alpha/2)},\ T^*_{(B(1-\alpha/2))} \right),$$

that is, just use the $\alpha/2$ and $1 - \alpha/2$ quantiles of the bootstrap sample. The justification for this interval is as follows. Suppose there exists a monotone transformation $U = m(T)$ such that $U \sim N(\phi, c^2)$ where $\phi = m(\theta)$. We do not suppose we know the transformation, only that one exists. Let $U^*_b = m(T^*_b)$. Note that $U^*_{(B\alpha/2)} = m(T^*_{(B\alpha/2)})$ since a monotone transformation preserves quantiles. Since, $U \sim N(\phi, c^2)$, the $\alpha/2$ quantile of U is $\phi - z_{\alpha/2}c$. Hence, $U^*_{(B\alpha/2)} = \phi - z_{\alpha/2}c \approx U - z_{\alpha/2}c$ and $U^*_{(B(1-\alpha/2))} \approx U + z_{\alpha/2}c$. Therefore,

$$
\begin{aligned}
\mathbb{P}(T^*_{B\alpha/2} \leq \theta \leq T^*_{B(1-\alpha/2)}) &= \mathbb{P}(m(T^*_{B\alpha/2}) \leq m(\theta) \leq m(T^*_{B(1-\alpha/2)})) \\
&= \mathbb{P}(U^*_{B\alpha/2} \leq \phi \leq U^*_{B(1-\alpha/2)}) \\
&\approx \mathbb{P}(U - cz_{\alpha/2} \leq \phi \leq U + cz_{\alpha/2}) \\
&= \mathbb{P}\left(-z_{\alpha/2} \leq \frac{U - \phi}{c} \leq z_{\alpha/2} \right) \\
&= 1 - \alpha.
\end{aligned}
$$

Amazingly, we never need to know m. Unfortunately, an exact normalizing transformation will rarely exist but there may exist approximate normalizing transformations. This has led to an industry of **adjusted percentile methods**, the most popular being the BC_a interval (bias-corrected and accelerated). We will not consider these intervals here.

3.17 Example. For estimating the skewness of the nerve data, here are the various confidence intervals.

Method	95% Interval
Normal	(1.44, 2.09)
percentile	(1.42, 2.03)
pivotal	(1.48, 2.11)
studentized	(1.45, 2.28)

The studentized interval requires some explanation. For each bootstrap replication we compute $\widehat{\theta}^*$ and we also need the standard error $\widehat{se}^*$ of $\widehat{\theta}^*$. We could do a bootstrap within the bootstrap (called a double bootstrap) but this is computationally expensive. Instead, we computed $\widehat{se}^*$ using the nonparametric delta method applied to the bootstrap sample as described in Example 3.10. ∎

3.5   Some Theory

Under certain conditions, $\widehat{G}_n^*$ is a consistent estimate of $G_n(t) = \mathbb{P}(T_n \leq t)$. To make this precise, let $\mathbb{P}_{\widehat{F}_n}(\cdot)$ denote probability statements made from $\widehat{F}_n$, treating the original data $X_1, \ldots, X_n$ as fixed. Assume that $T_n = T(\widehat{F}_n)$ is some functional of $\widehat{F}_n$. Then,

$$\widehat{G}_n^*(t) = \mathbb{P}_{\widehat{F}_n}(T(\widehat{F}_n^*) \leq t) = \mathbb{P}_{\widehat{F}_n}\left(\sqrt{n}(T(\widehat{F}_n^*) - T(F)) \leq u\right) \qquad (3.18)$$

where $u = \sqrt{n}(t - T(F))$. Consistency of the bootstrap can now be expressed as follows.

3.19 Theorem. *Suppose that* $\mathbb{E}(X_1^2) < \infty$. *Let* $T_n = g(\overline{X}_n)$ *where* g *is continuously differentiable at* $\mu = \mathbb{E}(X_1)$ *and that* $g'(\mu) \neq 0$. *Then,*

$$\sup_u \left| \mathbb{P}_{\widehat{F}_n}\left(\sqrt{n}(T(\widehat{F}_n^*) - T(\widehat{F}_n)) \leq u\right) - \mathbb{P}_F\left(\sqrt{n}(T(\widehat{F}_n) - T(F)) \leq u\right) \right| \overset{\text{a.s.}}{\longrightarrow} 0.$$
$$(3.20)$$

3.21 Theorem. *Suppose that* $T(F)$ *is Hadamard differentiable with respect to* $d(F, G) = \sup_x |F(x) - G(x)|$ *and that* $0 < \int L_F^2(x) dF(x) < \infty$. *Then,*

$$\sup_u \left| \mathbb{P}_{\widehat{F}_n}\left(\sqrt{n}(T(\widehat{F}_n^*) - T(\widehat{F}_n)) \leq u\right) - \mathbb{P}_F\left(\sqrt{n}(T(\widehat{F}_n) - T(F)) \leq u\right) \right| \overset{\text{P}}{\longrightarrow} 0.$$
$$(3.22)$$

Look closely at Theorems 3.19 and 3.21. It is because of results like these that the bootstrap "works." In particular, the validity of bootstrap confidence intervals depends on these theorems. See, for example, Theorem 3.16. There is a tendency to treat the bootstrap as a panacea for all problems. But the bootstrap requires regularity conditions to yield valid answers. It should not be applied blindly.

It can also be shown that the bootstrap variance estimate is consistent with some conditions on T. Generally, the conditions for consistency of the bootstrap are weaker than those for the jackknife. For example, the bootstrap estimate of the variance of the median is consistent, but the jackknife estimate of the variance of the median is not consistent (Theorem 3.7).

Let us now compare the accuracy of the different confidence interval methods. Consider a $1 - \alpha$ one-sided interval $[\widehat{\theta}_\alpha, \infty)$. We would like $\mathbb{P}(\theta \leq \widehat{\theta}_\alpha) = \alpha$ but usually this holds only approximately. If $\mathbb{P}(\theta \leq \widehat{\theta}_\alpha) = \alpha + O(n^{-1/2})$ then we say that the interval is first-order accurate. If $\mathbb{P}(\theta \leq \widehat{\theta}_\alpha) = \alpha + O(n^{-1})$ then we say that the interval is second-order accurate. Here is the comparison:

Method	Accuracy
Normal interval	first-order accurate
basic pivotal interval	first-order accurate
percentile interval	first-order accurate
studentized pivotal interval	second-order accurate
adjusted percentile interval	second-order accurate

Here is an explanation of why the studentized interval is more accurate. See Davison and Hinkley (1997), and Hall (1992a), for more details. Let $Z_n = \sqrt{n}(T_n - \theta)/\sigma$ be a standardized quantity that converges to a standard Normal. Thus $\mathbb{P}_F(Z_n \leq z) \to \Phi(z)$. In fact,

$$\mathbb{P}_F(Z_n \leq z) = \Phi(z) + \frac{1}{\sqrt{n}} a(z)\phi(z) + O\left(\frac{1}{n}\right) \tag{3.23}$$

for some polynomial a involving things like skewness. The bootstrap version satisfies

$$\mathbb{P}_{\widehat{F}}(Z_n^* \leq z) = \Phi(z) + \frac{1}{\sqrt{n}} \widehat{a}(z)\phi(z) + O_P\left(\frac{1}{n}\right) \tag{3.24}$$

where $\widehat{a}(z) - a(z) = O_P(n^{-1/2})$. Subtracting, we get

$$\mathbb{P}_F(Z_n \leq z) - \mathbb{P}_{\widehat{F}}(Z_n^* \leq z) = O_P\left(\frac{1}{n}\right). \tag{3.25}$$

Now suppose we work with the nonstudentized quantity $V_n = \sqrt{n}(T_n - \theta)$. Then,

$$\mathbb{P}_F(V_n \leq z) = \mathbb{P}_F\left(\frac{V_n}{\sigma} \leq \frac{z}{\sigma}\right)$$

$$= \Phi\left(\frac{z}{\sigma}\right) + \frac{1}{\sqrt{n}} b\left(\frac{z}{\sigma}\right) \phi\left(\frac{z}{\sigma}\right) + O\left(\frac{1}{n}\right)$$

for some polynomial b. For the bootstrap we have

$$
\begin{aligned}
\mathbb{P}_{\widehat{F}}(V_n^* \leq z) &= \mathbb{P}_{\widehat{F}}\left(\frac{V_n}{\widehat{\sigma}} \leq \frac{z}{\widehat{\sigma}}\right) \\
&= \Phi\left(\frac{z}{\widehat{\sigma}}\right) + \frac{1}{\sqrt{n}}\widehat{b}\left(\frac{z}{\widehat{\sigma}}\right)\phi\left(\frac{z}{\widehat{\sigma}}\right) + O_P\left(\frac{1}{n}\right)
\end{aligned}
$$

where $\widehat{\sigma} = \sigma + O_P(n^{-1/2})$. Subtracting, we get

$$
\mathbb{P}_F(V_n \leq z) - \mathbb{P}_{\widehat{F}}(V_n^* \leq z) = O_P\left(\frac{1}{\sqrt{n}}\right) \tag{3.26}
$$

which is less accurate than (3.25).

3.6   Bibliographic Remarks

The jackknife was invented by Quenouille (1949) and Tukey (1958). The bootstrap was invented by Efron (1979). There are several books on these topics including Efron and Tibshirani (1993), Davison and Hinkley (1997), Hall (1992a) and Shao and Tu (1995). Also, see Section 3.6 of van der Vaart and Wellner (1996).

3.7   Appendix

The book by Shao and Tu (1995) gives an explanation of the techniques for proving consistency of the jackknife and the bootstrap. Following Section 3.1 of their text, let us look at two ways of showing that the bootstrap is consistent for the case $T_n = \overline{X}_n = n^{-1}\sum_{i=1}^n X_i$. Let $X_1, \ldots, X_n \sim F$ and let $T_n = \sqrt{n}(\overline{X}_n - \mu)$ where $\mu = \mathbb{E}(X_1)$. Let $H_n(t) = \mathbb{P}_F(T_n \leq t)$ and let $\widehat{H}_n(t) = \mathbb{P}_{\widehat{F}_n}(T_n^* \leq t)$ be the bootstrap estimate of H_n where $T_n^* = \sqrt{n}(\overline{X}_n^* - \overline{X}_n)$ and $X_1^*, \ldots, X_n^* \sim \widehat{F}_n$. Our goal is to show that $\sup_x |H_n(x) - \widehat{H}_n(x)| \xrightarrow{\text{a.s.}} 0$.

The first method was used by Bickel and Freedman (1981) and is based on Mallows' metric. If X and Y are random variables with distributions F and G, Mallows' metric is defined by $d_r(F, G) = d_r(X, Y) = \inf\left(\mathbb{E}|X - Y|^r\right)^{1/r}$ where the infimum is over all joint distributions with marginals F and G. Here are some facts about d_r. Let $X_n \sim F_n$ and $X \sim F$. Then, $d_r(F_n, F) \to 0$ if and only if $X_n \rightsquigarrow X$ and $\int |x|^r dF_n(x) \to \int |x|^r dF(x)$. If $\mathbb{E}(|X_1|^r) < \infty$ then $d_r(\widehat{F}_n, F) \xrightarrow{\text{a.s.}} 0$. For any constant a, $d_r(aX, aY) = |a|d_r(X, Y)$. If $\mathbb{E}(X^2) < \infty$

and $\mathbb{E}(Y^2) < \infty$ then $d_2(X, Y)^2 = \big(d_2(X - \mathbb{E}(X), Y - \mathbb{E}(Y))\big)^2 + |\mathbb{E}(X - Y)|^2$. If $\mathbb{E}(X_j) = \mathbb{E}(Y_j)$ and $\mathbb{E}(|X_j|^r) < \infty$, $\mathbb{E}(|Y_j|^r) < \infty$ then

$$\left(d_2\left(\sum_{j=1}^m X_j, \sum_{j=1}^m Y_j\right)\right)^2 \le \sum_{j=1}^m d_2(X_j, Y_j)^2.$$

Using the properties of d_r we have

$$
\begin{aligned}
d_2(\widehat{H}_n, H_n) &= d_2(\sqrt{n}(\overline{X}_n^* - \overline{X}_n), \sqrt{n}(\overline{X}_n - \mu)) \\
&= \frac{1}{\sqrt{n}} d_2\left(\sum_{i=1}^n (X_i^* - \overline{X}_n), \sum_{i=1}^n (X_i - \mu)\right) \\
&\le \sqrt{\frac{1}{n} \sum_{i=1}^n d_2(X_i^* - \overline{X}_n, X_i - \mu)^2} \\
&= d_2(X_1^* - X_1, X_1 - \mu) \\
&= \sqrt{d_2(X_1^*, X_1)^2 - (\mu - \mathbb{E}_* X_1^*)^2} \\
&= \sqrt{d_2(\widehat{F}_n, F)^2 - (\mu - \overline{X}_n)^2} \\
&\xrightarrow{\text{a.s.}} 0
\end{aligned}
$$

since $d_2(\widehat{F}_n, F) \xrightarrow{\text{a.s.}} 0$ and $\overline{X}_n \xrightarrow{\text{a.s.}} \mu$. Hence, $\sup_x |H_n(x) - \widehat{H}_n(x)| \xrightarrow{\text{a.s.}} 0$.

The second method, due to Singh (1981), uses the Berry–Esséen bound (1.28) which we now review. Let $X_1, \ldots, X_n$ be IID with finite mean $\mu = \mathbb{E}(X_1)$, variance $\sigma^2 = \mathbb{V}(X_1)$ and third moment, $\mathbb{E}|X_1|^3 < \infty$. Let $Z_n = \sqrt{n}(\overline{X}_n - \mu)/\sigma$. Then

$$\sup_z |\mathbb{P}(Z_n \le z) - \Phi(z)| \le \frac{33}{4} \frac{\mathbb{E}|X_1 - \mu|^3}{\sqrt{n}\sigma^3}. \tag{3.27}$$

Let $Z_n^* = (\overline{X}_n^* - \overline{X}_n)/\widehat{\sigma}$ where $\widehat{\sigma}^2 = n^{-1} \sum_{i=1}^n (X_i - \overline{X}_n)^2$. Replacing F with $\widehat{F}_n$ and $\overline{X}_n$ with $\overline{X}_n^*$ we get

$$\sup_z |\mathbb{P}_{\widehat{F}_n}(Z_n^* \le z) - \Phi(z)| \le \frac{33}{4} \frac{\sum_{i=1}^n |X_i - \overline{X}_n|^3}{n^{3/2}\widehat{\sigma}^3}. \tag{3.28}$$

Let $d(F, G) = \sup_x |F(x) - G(x)|$ and define $\Phi_a(x) = \Phi(x/a)$. Then

$$
\begin{aligned}
\sup_z |\mathbb{P}_{\widehat{F}_n}(Z_n^* \le z) - \Phi(z)| &= \sup_z \left| \mathbb{P}_{\widehat{F}_n}\left(\sqrt{n}(\overline{X}_n^* - \overline{X}_n) \le z\widehat{\sigma}\right) - \Phi\left(\frac{z\widehat{\sigma}}{\widehat{\sigma}}\right) \right| \\
&= \sup_t \left| \mathbb{P}_{\widehat{F}_n}\left(\sqrt{n}(\overline{X}_n^* - \overline{X}_n) \le t\right) - \Phi_{\widehat{\sigma}}(t) \right| \\
&= d(\widehat{H}_n, \Phi_{\widehat{\sigma}}).
\end{aligned}
$$

By the triangle inequality

$$d(\widehat{H}_n, H_n) \leq d(\widehat{H}_n, \Phi_{\widehat{\sigma}}) + d(\Phi_{\widehat{\sigma}}, \Phi_\sigma) + d(\Phi_\sigma, H_n). \tag{3.29}$$

The third term in (3.29) goes to 0 by the central limit theorem. For the second term, $d(\Phi_{\widehat{\sigma}}, \Phi_\sigma) \xrightarrow{\text{a.s.}} 0$ since $\widehat{\sigma}^2 \xrightarrow{\text{a.s.}} \sigma^2 = \mathbb{V}(X_1)$. The first term is bounded by the right-hand side of (3.28). Since $\mathbb{E}(X_1^2) < \infty$, this goes to 0 by the following result: if $\mathbb{E}|X_1|^\delta < \infty$ for some $0 < \delta < 1$ then $n^{-1/\delta} \sum_{i=1}^n |X_i| \xrightarrow{\text{a.s.}} 0$. In conclusion, $d(\widehat{H}_n, H_n) \xrightarrow{\text{a.s.}} 0$.

3.8   Exercises

1. Let $T(F) = \int (x - \mu)^3 dF(x)/\sigma^3$ be the skewness. Find the influence function.

2. The following data were used to illustrate the bootstrap by Bradley Efron, the inventor of the bootstrap. The data are LSAT scores (for entrance to law school) and GPA.

LSAT	576	635	558	578	666	580	555	661
	651	605	653	575	545	572	594	

GPA	3.39	3.30	2.81	3.03	3.44	3.07	3.00	3.43
	3.36	3.13	3.12	2.74	2.76	2.88	3.96	

Each data point is of the form $X_i = (Y_i, Z_i)$ where $Y_i = \text{LSAT}_i$ and $Z_i = \text{GPA}_i$. Find the plug-in estimate of the correlation coefficient. Estimate the standard error using (i) the influence function, (ii) the jackknife and (iii) the bootstrap. Next compute a 95 percent studentized pivotal bootstrap confidence interval. You will need to compute the standard error of T^* for every bootstrap sample.

3. Let $T_n = \overline{X}_n^2$, $\mu = \mathbb{E}(X_1)$, $\alpha_k = \int |x - \mu|^k dF(x)$ and $\widehat{\alpha}_k = n^{-1} \sum_{i=1}^n |X_i - \overline{X}_n|^k$. Show that

$$v_{\text{boot}} = \frac{4\overline{X}_n^2 \widehat{\alpha}_2}{n} + \frac{4\overline{X}_n \widehat{\alpha}_3}{n^2} + \frac{\widehat{\alpha}_4}{n^3}.$$

4. Prove Theorem 3.16.

5. Repeat the calculations in Example 3.17 but use a parametric bootstrap. Assume that the data are log-Normal. That is, assume that $Y \sim N(\mu, \sigma^2)$ where $Y = \log X$. You will draw bootstrap samples $Y_1^*, \ldots, Y_n^*$ from $N(\hat{\mu}, \hat{\sigma}^2)$. Then set $X_i^* = e^{Y_i^*}$.

6. (Computer experiment.) Conduct a simulation to compare the four bootstrap confidence interval methods. Let $n = 50$ and let $T(F) = \int (x - \mu)^3 dF(x)/\sigma^3$ be the skewness. Draw $Y_1, \ldots, Y_n \sim N(0, 1)$ and set $X_i = e^{Y_i}$, $i = 1, \ldots, n$. Construct the four types of bootstrap 95 percent intervals for $T(F)$ from the data $X_1, \ldots, X_n$. Repeat this whole thing many times and estimate the true coverage of the four intervals.

7. Let
$$X_1, \ldots, X_n \sim t_3$$
where $n = 25$. Let $\theta = T(F) = (q_{.75} - q_{.25})/1.34$ where q_p denotes the pth quantile. Do a simulation to compare the coverage and length of the following confidence intervals for θ: (i) Normal interval with standard error from the jackknife, (ii) Normal interval with standard error from the bootstrap, (iii) bootstrap percentile interval.

   Remark: The jackknife does not give a consistent estimator of the variance of a quantile.

8. Let $X_1, \ldots, X_n$ be distinct observations (no ties). Show that there are
$$\binom{2n-1}{n}$$
distinct bootstrap samples.

   Hint: Imagine putting n balls into n buckets.

9. Let $X_1, \ldots, X_n$ be distinct observations (no ties). Let $X_1^*, \ldots, X_n^*$ denote a bootstrap sample and let $\overline{X}_n^* = n^{-1} \sum_{i=1}^n X_i^*$. Find: $\mathbb{E}(\overline{X}_n^* | X_1, \ldots, X_n)$, $\mathbb{V}(\overline{X}_n^* | X_1, \ldots, X_n)$, $\mathbb{E}(\overline{X}_n^*)$ and $\mathbb{V}(\overline{X}_n^*)$.

10. (Computer experiment.) Let $X_1, \ldots, X_n \sim \text{Normal}(\mu, 1)$. Let $\theta = e^\mu$ and let $\hat{\theta} = e^{\overline{X}}$ be the MLE. Create a data set (using $\mu = 5$) consisting of n=100 observations.

    (a) Use the delta method to get the se and 95 percent confidence interval for θ. Use the parametric bootstrap to get the se and 95 percent confidence interval for θ. Use the nonparametric bootstrap to get the se and 95 percent confidence interval for θ. Compare your answers.

(b) Plot a histogram of the bootstrap replications for the parametric and nonparametric bootstraps. These are estimates of the distribution of $\widehat{\theta}$. The delta method also gives an approximation to this distribution, namely, Normal$(\widehat{\theta}, \widehat{se}^2)$. Compare these to the true sampling distribution of $\widehat{\theta}$. Which approximation—parametric bootstrap, bootstrap or delta method—is closer to the true distribution?

11. Let $X_1, \ldots, X_n \sim$ Uniform$(0, \theta)$. The MLE is

$$\widehat{\theta} = X_{\max} = \max\{X_1, \ldots, X_n\}.$$

Generate a data set of size 50 with $\theta = 1$.

(a) Find the distribution of $\widehat{\theta}$. Compare the true distribution of $\widehat{\theta}$ to the histograms from the parametric and nonparametric bootstraps.

(b) This is a case where the nonparametric bootstrap does very poorly. In fact, we can prove that this is the case. Show that, for the parametric bootstrap $\mathbb{P}(\widehat{\theta}^* = \widehat{\theta}) = 0$ but for the nonparametric bootstrap $\mathbb{P}(\widehat{\theta}^* = \widehat{\theta}) \approx .632$. *Hint*: Show that, $P(\widehat{\theta}^* = \widehat{\theta}) = 1 - (1 - (1/n))^n$. Then take the limit as n gets large.

12. Suppose that 50 people are given a placebo and 50 are given a new treatment. Thirty placebo patients show improvement, while 40 treated patients show improvement. Let $\tau = p_2 - p_1$ where p_2 is the probability of improving under treatment and p_1 is the probability of improving under placebo.

(a) Find the MLE of τ. Find the standard error and 90 percent confidence interval using the delta method.

(b) Find the standard error and 90 percent confidence interval using the bootstrap.

13. Let $X_1, \ldots, X_n \sim F$ be IID and let $X_1^*, \ldots, X_n^*$ be a bootstrap sample from $\widehat{F}_n$. Let G denote the marginal distribution of X_i^*. Note that $G(x) = \mathbb{P}(X_i^* \leq x) = \mathbb{E}\mathbb{P}(X_i^* \leq x | X_1, \ldots, X_n) = \mathbb{E}(\widehat{F}_n(x)) = F(x)$. So it appears that X_i^* and X_i have the same distribution. But in Exercise 9 we showed that $\mathbb{V}(\overline{X}_n) \neq \mathbb{V}(\overline{X}_n^*)$. This appears to be a contradiction. Explain.

4
Smoothing: General Concepts

To estimate a curve—such as a probability density function f or a regression function r—we must smooth the data in some way. The rest of the book is devoted to smoothing methods. In this chapter we discuss some general issues related to smoothing. There are mainly two types of problems we will study. The first is **density estimation** in which we have a sample $X_1, \ldots, X_n$ from a distribution F with density f, written

$$X_1, \ldots, X_n \sim f, \tag{4.1}$$

and we want to estimate the probability density function f. The second is **regression** in which we have a pairs $(x_1, Y_1), \ldots, (x_n, Y_n)$ where

$$Y_i = r(x_i) + \epsilon_i \tag{4.2}$$

where $\mathbb{E}(\epsilon_i) = 0$, and we want to estimate the regression function r. We begin with some examples all of which are discussed in more detail in the chapters that follow.

4.3 Example (Density estimation). Figure 4.1 shows histograms of 1266 data points from the Sloan Digital Sky Survey (SDSS). As described on the SDSS website `www.sdss.org`:

> Simply put, the Sloan Digital Sky Survey is the most ambitious astronomical survey project ever undertaken. The survey will map

in detail one-quarter of the entire sky, determining the positions and absolute brightnesses of more than 100 million celestial objects. It will also measure the distances to more than a million galaxies and quasars.

Each data point X_i is a redshift,[1] essentially the distance from us to a galaxy. The data lie on a "pencil beam" which means that the sample lies in a narrow tube starting at the earth and pointing out into space; see Figure 4.2. The full dataset is three-dimensional. We have extracted data along this pencil beam to make it one-dimensional. Our goal is to understand the distribution of galaxies. In particular, astronomers are interested in finding clusters of galaxies. Because the speed of light is finite, when we look at galaxies that are farther away we are seeing further back in time. By looking at how galaxies cluster as a function of redshift, we are seeing how galaxy clustering evolved over time.

We regard the redshifts $X_1, \ldots, X_n$ as a sample from a distribution F with density f, that is,

$$X_1, \ldots, X_n \sim f$$

as in (4.1). One way to find galaxy clusters is to look for peaks in the density f. A histogram is a simple method for estimating the density f. The details will be given in Chapter 6 but here is a brief description. We cut the real line into intervals, or bins, and count the number of observations in each bin. The heights of the bars in the histogram are proportional to the count in each bin. The three histograms in Figure 4.1 are based on different numbers of bins. The top left histogram uses a large number of bins, the top right histogram uses fewer bins and the bottom left histogram uses fewer still. The width h of the bins is a **smoothing parameter**. We will see that large h (few bins) leads to an estimator with large bias but small variance, called **oversmoothing**. Small h (many bins) leads to an estimator with small bias but large variance, called **undersmoothing**. The bottom right plot shows an estimate of the **mean squared error** (MSE) of the histogram estimator, which is a measure of the inaccuracy of the estimator. The estimated MSE is shown as a function of the number of bins. The top right histogram has 308 bins, which corresponds to the minimizer of the estimated MSE.

---

[1] When an object moves away from us, its light gets shifted towards the red end of the spectrum, called redshift. The faster an object moves away from us, the more its light is redshifted. Objects further away are moving away from us faster than close objects. Hence the redshift can be used to deduce the distance. This is more complicated than it sounds because the conversion from redshift to distance requires knowledge of the geometry of the universe.

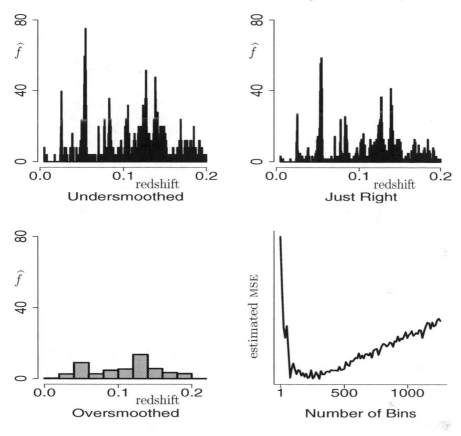

FIGURE 4.1. Three histograms for the astronomy data. The top left histogram has too many bins. The bottom left histogram has too few bins. The top right histogram uses 308 bins, chosen by the method of cross-validation as described in Chapter 6. The lower right plot shows the estimated mean squared error (inaccuracy) versus the number of bins.

Figure 4.3 shows a more sophisticated estimator of f called a **kernel estimator** which we will describe in Chapter 6. Again there is a smoothing parameter h and the three estimators correspond to increasing h. The structure in the data is evident only if we smooth the data by just the right amount (top right plot). ∎

4.4 Example (Nonparametric regression). The beginning of the universe is usually called the Big Bang. It is misleading to think of this as an event taking place in empty space. More accurately, the early universe was in a hot, dense state. Since then, the universe has expanded and cooled. The remnants of the heat from the Big Bang are still observable and are called the cosmic mi-

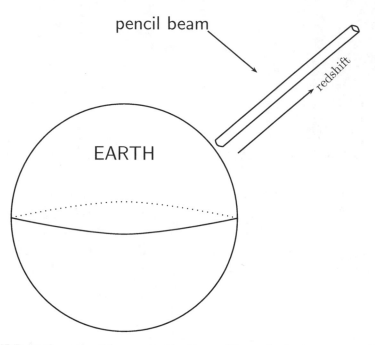

FIGURE 4.2. In a pencil beam sample, the positions of galaxies are recorded along a path from the Earth and outward.

crowave background (CMB) radiation. Figure 4.4 shows CMB data from the Wilkinson Microwave Anisotropy Probe (WMAP). Information on WMAP is available at http://map.gsfc.nasa.gov. The image shows the temperature at each point in the sky. This is a snapshot of the universe 379,000 years after the Big Bang. The average temperature is 2.73 degrees above absolute zero but the temperature is not constant across the sky. The fluctuations in the temperature map provide information about the early universe. Indeed, as the universe expanded, there was a tug of war between the force of expansion and contraction due to gravity. This caused acoustic waves in the hot gas (like a vibrating bowl of jello), which is why there are temperature fluctuations. The strength of the temperature fluctuations $r(x)$ at each frequency (or multipole) x is called the power spectrum and this power spectrum can be used to by cosmologists to answer cosmological questions (Genovese et al. (2004)). For example, the relative abundance of different constituents of the universe (such as baryons and dark matter) correspond to peaks in the power spectrum. Through a very complicated procedure that we won't describe here, the temperature map can be reduced to a scatterplot of power versus frequency.

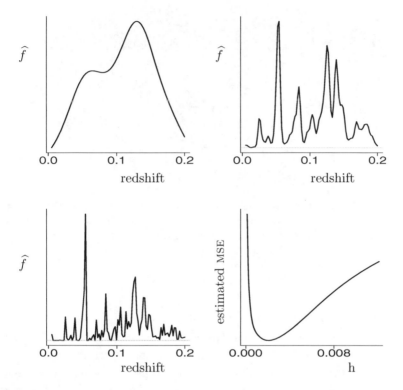

FIGURE 4.3. Kernel density estimators and estimated MSE for the astronomy data. Top left: oversmoothed. Top right: just right (bandwidth chosen by cross-validation). Bottom left: undersmoothed. Bottom right: estimated MSE as a function of the smoothing parameter h.

The first 400 data points are shown in Figure 4.5. (All 899 data points are shown in Figure 5.3.)

At this point of the analysis, the data consist of n pairs $(x_1, Y_1), \ldots, (x_n, Y_n)$ where x_i is called the multipole moment and Y_i is the estimated power spectrum of the temperature fluctuations. If $r(x)$ denotes the true power spectrum then

$$Y_i = r(x_i) + \epsilon_i$$

where ϵ_i is a random error with mean 0, as in (4.2). The goal in nonparametric regression is to estimate r making only minimal assumptions about r. The first plot in Figure 4.5 shows the data and the next three plots show a nonparametric estimator (called a local regression estimator) of r with increasing smoothing parameter h. The structure of the data is distorted if we smooth too little or too much. The details will be explained in the next chapter. ∎

FIGURE 4.4. The WMAP (Wilkinson Microwave Anisotropy Probe) temperature map. This is the heat left over from the Big Bang. The data show the temperature at each point in the sky. The microwave light captured in this picture is from 379,000 years after the Big Bang (13 billion years ago). The average temperature is 2.73 degrees above absolute zero. The fluctuations in the temperature map provide important information about the early universe.

4.5 Example (Nonparametric regression). Ruppert et al. (2003) describe data from a light detection and ranging (LIDAR) experiment. LIDAR is used to monitor pollutants; see Sigrist (1994). Figure 4.6 shows 221 observations. The response is the log of the ratio of light received from two lasers. The frequency of one laser is the resonance frequency of mercury while the second has a different frequency. The estimates shown here are called **regressograms**, which is the regression version of a histogram. We divide the horizontal axis into bins and then we take the sample average of the Y_is in each bin. The smoothing parameter h is the width of the bins. As the binsize h decreases, the estimated regression function $\widehat{r}_n$ goes from oversmoothing to undersmoothing. ∎

4.6 Example (Nonparametric binary regression). This example is from Pagano and Gauvreau (1993); it also appears in Ruppert et al. (2003). The goal is to relate the presence or absence of bronchopulmonary dysplasia (BPD) with birth weight (in grams) for 223 babies. BPD is a chronic lung disease that can affect premature babies. The outcome Y is binary: $Y = 1$ if the baby has BPD and $Y = 0$ otherwise. The covariate is $x =$ birth weight. A common parametric model for relating a binary outcome Y to a covariate x is **logistic regression** which has the form

$$r(x; \beta_0, \beta_1) \equiv \mathbb{P}(Y = 1 | X = x) = \frac{e^{\beta_0 + \beta_1 x}}{1 + e^{\beta_0 + \beta_1 x}}.$$

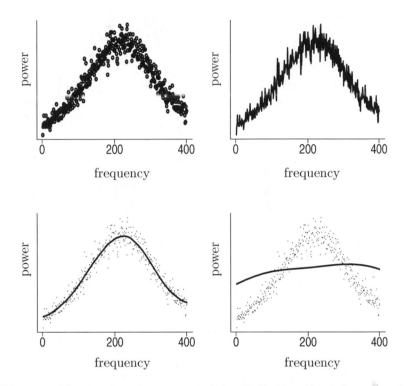

FIGURE 4.5. The first 400 data points of the CMB data. Top left: scatterplot of power versus frequency. Top right: undersmoothed. Bottom left: just right; Bottom right: oversmoothed.

The parameters β_0 and β_1 are usually estimated by maximum likelihood. The estimated function (solid line) $r(x; \widehat{\beta}_0, \widehat{\beta}_1)$ together with the data are shown in Figure 4.7. Also shown are two nonparametric estimates. In this example, the nonparametric estimators are not terribly different from the parametric estimator. Of course, this need not always be the case. ∎

4.7 Example (Multiple nonparametric regression). This example, from Venables and Ripley (2002), involves three covariates and one response variable. The data are 48 rock samples from a petroleum reservoir. The response is permeability (in milli-Darcies). The covariates are: the area of pores (in pixels out of 256 by 256), perimeter in pixels and shape (perimeter/$\sqrt{\text{area}}$). The goal is to predict permeability from the three covariates. One nonparametric model is

$$\text{permeability} = r(\text{area}, \text{perimeter}, \text{shape}) + \epsilon$$

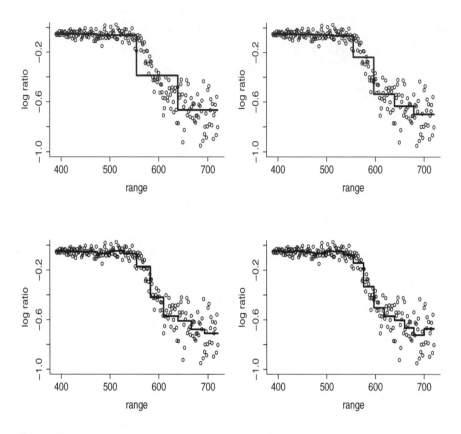

FIGURE 4.6. The LIDAR data from Example 4.5. The estimates are regressograms, obtained by averaging the Y_is over bins. As we decrease the binwidth h, the estimator becomes less smooth.

where r is a smooth function. A simpler, but less general model is the **additive model**

$$\text{permeability} = r_1(\text{area}) + r_2(\text{perimeter}) + r_3(\text{shape}) + \epsilon$$

for smooth functions r_1, r_2, and r_3. Estimates of r_1, r_2 and r_3 are shown in Figure 4.8. ∎

4.1 The Bias–Variance Tradeoff

Let $\widehat{f}_n(x)$ be an estimate of a function $f(x)$. The **squared error** (or L_2) loss function is

$$L\big(f(x), \widehat{f}_n(x)\big) = \big(f(x) - \widehat{f}_n(x)\big)^2. \tag{4.8}$$

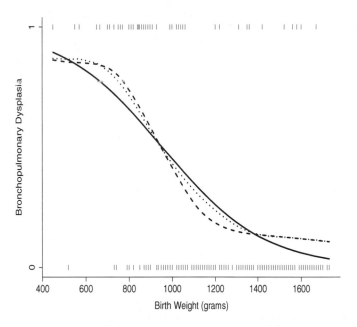

FIGURE 4.7. The BPD data from Example 4.6. The data are shown with small vertical lines. The estimates are from logistic regression (solid line), local likelihood (dashed line) and local linear regression (dotted line).

The average of this loss is called the **risk** or **mean squared error** (MSE) and is denoted by:

$$\text{MSE} = R(f(x), \widehat{f}_n(x)) = \mathbb{E}\big(L(f(x), \widehat{f}_n(x))\big). \tag{4.9}$$

The random variable in equation (4.9) is the function $\widehat{f}_n$ which implicitly depends on the observed data. We will use the terms risk and MSE interchangeably. A simple calculation (Exercise 2) shows that

$$R\big(f(x), \widehat{f}_n(x)\big) = \text{bias}_x^2 + \mathbb{V}_x \tag{4.10}$$

where

$$\text{bias}_x = \mathbb{E}(\widehat{f}_n(x)) - f(x)$$

is the bias of $\widehat{f}_n(x)$ and

$$\mathbb{V}_x = \mathbb{V}(\widehat{f}_n(x))$$

is the variance of $\widehat{f}_n(x)$. In words:

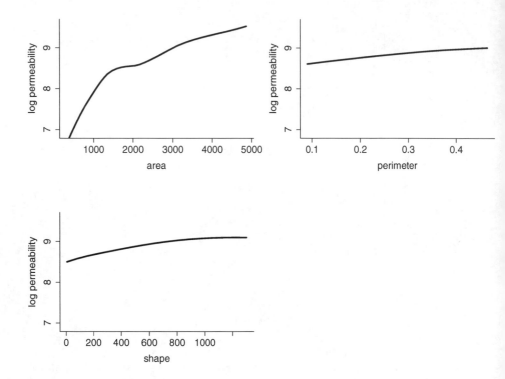

FIGURE 4.8. The rock data from Example 4.7. The plots show $\widehat{r}_1$, $\widehat{r}_2$, and $\widehat{r}_3$ for the additive model $Y = \widehat{r}_1(x_1) + \widehat{r}_2(x_2) + \widehat{r}_3(x_3) + \epsilon$.

$$\text{RISK} = \text{MSE} = \text{BIAS}^2 + \text{VARIANCE}. \tag{4.11}$$

The above definitions refer to the risk at a point x. Now we want to summarize the risk over different values of x. In density estimation problems, we will use the **integrated risk** or **integrated mean squared error** defined by

$$R(f, \widehat{f}_n) = \int R(f(x), \widehat{f}_n(x))dx. \tag{4.12}$$

For regression problems we can use the integrated MSE or the **average mean squared error**

$$R(r, \widehat{r}_n) = \frac{1}{n} \sum_{i=1}^{n} R(r(x_i), \widehat{r}_n(x_i)). \tag{4.13}$$

The average risk has the following predictive risk interpretation. Suppose the model for the data is the nonparametric regression model

$$Y_i = r(x_i) + \epsilon_i.$$

Suppose we draw a new observation $Y_i^* = r(x_i) + \epsilon_i^*$ at each x_i. If we predict Y_i^* with $\widehat{r}_n(x_i)$ then the **squared prediction error** is

$$(Y_i^* - \widehat{r}_n(x_i))^2 = (r(x_i) + \epsilon_i^* - \widehat{r}_n(x_i))^2.$$

Define the **predictive risk**

$$\textbf{predictive risk} = \mathbb{E}\left(\frac{1}{n}\sum_{i=1}^n (Y_i^* - \widehat{r}_n(x_i))^2\right).$$

Then, we have

$$\text{predictive risk} = R(r, \widehat{r}_n) + c \tag{4.14}$$

where $c = n^{-1}\sum_{i=1}^n \mathbb{E}((\epsilon_i^*)^2)$ is a constant. In particular, if each ϵ_i has variance σ^2 then

$$\text{predictive risk} = R(r, \widehat{r}_n) + \sigma^2. \tag{4.15}$$

Thus, up to a constant, the average risk and the predictive risk are the same.

The main challenge in smoothing is to determine how much smoothing to do. When the data are oversmoothed, the bias term is large and the variance is small. When the data are undersmoothed the opposite is true; see Figure 4.9. This is called the **bias–variance tradeoff**. Minimizing risk corresponds to balancing bias and variance.

4.16 Example. To understand the bias–variance tradeoff better, let f be a PDF and consider estimating $f(0)$. Let h be a small, positive number. Define

$$p_h \equiv \mathbb{P}\left(-\frac{h}{2} < X < \frac{h}{2}\right) = \int_{-h/2}^{h/2} f(x)dx \approx hf(0)$$

and hence

$$f(0) \approx \frac{p_h}{h}.$$

Let X be the number of observations in the interval $(-h/2, h/2)$. Then $X \sim$ Binomial(n, p_h). An estimate of p_h is $\widehat{p}_h = X/n$ and hence an estimate of $f(0)$ is

$$\widehat{f}_n(0) = \frac{\widehat{p}_h}{h} = \frac{X}{nh}. \tag{4.17}$$

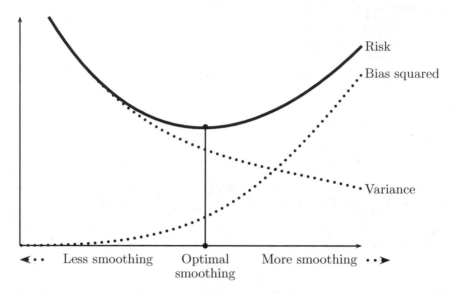

FIGURE 4.9. The bias–variance tradeoff. The bias increases and the variance decreases with the amount of smoothing. The optimal amount of smoothing, indicated by the vertical line, minimizes the risk = bias^2 + variance.

We will now show that the MSE of this estimator takes the form

$$\text{MSE} \approx Ah^4 + \frac{B}{nh} \tag{4.18}$$

for some constants A and B. The first term corresponds to the squared bias and the second term corresponds to the variance.

Since X is Binomial, it has mean np_h. Now,

$$f(x) \approx f(0) + xf'(0) + \frac{x^2}{2}f''(0).$$

So

$$p_h = \int_{-h/2}^{h/2} f(x)dx \approx \int_{-h/2}^{h/2}\left(f(0) + xf'(0) + \frac{x^2}{2}f''(0)\right) dx = hf(0) + \frac{f''(0)h^3}{24}$$

and hence, from (4.17),

$$\mathbb{E}(\widehat{f}_n(0)) = \frac{\mathbb{E}(X)}{nh} = \frac{p_h}{h} \approx f(0) + \frac{f''(0)h^2}{24}.$$

Therefore, the bias is

$$\text{bias} = \mathbb{E}(\widehat{f}_n(0)) - f(0) \approx \frac{f''(0)h^2}{24}. \tag{4.19}$$

To compute the variance, note that $\mathbb{V}(X) = np_h(1 - p_h)$. Therefore,

$$\mathbb{V}(\widehat{f}_n(0)) = \frac{\mathbb{V}(X)}{n^2 h^2} = \frac{p_h(1 - p_h)}{nh^2} \approx \frac{p_h}{nh^2}$$

where we have used the fact that $1 - p_h \approx 1$ since h is small. Thus,

$$\mathbb{V}(\widehat{f}_n(0)) \approx \frac{hf(0) + \frac{f''(0)h^3}{24}}{nh^2} = \frac{f(0)}{nh} + \frac{f''(0)h}{24n} \approx \frac{f(0)}{nh}. \tag{4.20}$$

Therefore,

$$\text{MSE} = \text{bias}^2 + \mathbb{V}(\widehat{f}_n(0)) \approx \frac{(f''(0))^2 h^4}{576} + \frac{f(0)}{nh} \equiv Ah^4 + \frac{B}{nh}. \tag{4.21}$$

As we smooth more, that is, as we increase h, the bias term increases and the variance term decreases. As we smooth less, that is, as we decrease h, the bias term decreases and the variance term increases. This is a typical bias–variance analysis. ∎

4.2   Kernels

Throughout the rest of the book, we will often use the word "kernel." For our purposes, the word **kernel** refers to any smooth function K such that $K(x) \geq 0$ and

$$\int K(x)\,dx = 1, \quad \int xK(x)dx = 0 \quad \text{and} \quad \sigma_K^2 \equiv \int x^2 K(x)dx > 0. \tag{4.22}$$

Some commonly used kernels are the following:

$$\text{the boxcar kernel}: \quad K(x) = \frac{1}{2}I(x),$$

$$\text{the Gaussian kernel}: \quad K(x) = \frac{1}{\sqrt{2\pi}}e^{-x^2/2},$$

$$\text{the Epanechnikov kernel}: \quad K(x) = \frac{3}{4}(1 - x^2)I(x)$$

$$\text{the tricube kernel}: \quad K(x) = \frac{70}{81}(1 - |x|^3)^3 I(x)$$

where

$$I(x) = \begin{cases} 1 & \text{if } |x| \leq 1 \\ 0 & \text{if } |x| > 1. \end{cases}$$

These kernels are plotted in Figure 4.10.

Kernels are used for taking local averages. For example, suppose we have pairs of data $(x_1, Y_1), \ldots, (x_n, Y_n)$ and we want to take the average of all the

Y_i's whose corresponding x_i's are within a distance h of some point x. This local average is equal to

$$\sum_{i=1}^{n} Y_i \, \ell_i(x) \tag{4.23}$$

where

$$\ell_i(x) = \frac{K\left(\frac{X_i - x}{h}\right)}{\sum_{i=1}^{n} K\left(\frac{X_i - x}{h}\right)} \tag{4.24}$$

and K is the boxcar kernel. If we replace the boxcar kernel with another kernel, then (4.23) becomes a locally weighted average. Kernels will play an important role in many estimation methods. The smoother kernels lead to smoother estimates and are usually preferred to the boxcar kernel.

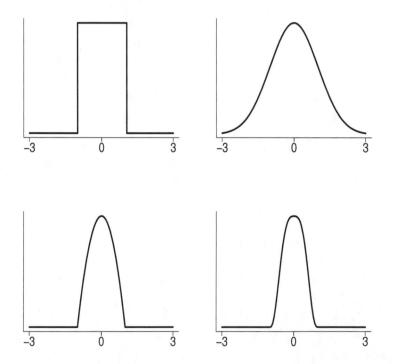

FIGURE 4.10. Examples of kernels: boxcar (top left), Gaussian (top right), Epanechnikov (bottom left), and tricube (bottom right).

4.3 Which Loss Function?

We can use other loss functions besides squared error. The L_p loss

$$\left\{ \int |f(x) - \widehat{f}_n(x)|^p \right\}^{1/p}$$

has received some attention especially L_1 which tends to be outlier resistant and is invariant under one-to-one transformations. The results for other L_p loss functions are not dramatically different than L_2 but L_p is much harder to work with when $p \neq 2$. Hence, L_2 remains the popular choice. In the machine learning community there is some interest in Kullback–Leibler loss

$$L(f, \widehat{f}_n) = \int f(x) \left(\log f(x)/\widehat{f}_n(x) \right) dx.$$

In fact, maximum likelihood estimation implicitly uses this loss function. Despite the natural role of Kullback–Leibler distance in parametric statistics, it is usually not an appropriate loss function in smoothing problems because it is exquisitely sensitive to the tails of the distribution; see Hall (1987). As a result, the tails can drive the whole estimation process.

4.4 Confidence Sets

Providing an estimate $\widehat{f}_n$ of a curve f is rarely sufficient for drawing scientific conclusions. In the chapters that follow we shall also provide a confidence set for f. This will either take the form of a ball

$$\mathcal{B}_n = \left\{ f : \int (f(x) - \widehat{f}_n(x))^2 \, dx \leq s_n^2 \right\}$$

with some radius s_n or a band (or envelope)

$$\mathcal{B}_n = \left\{ f : \ell(x) \leq f(x) \leq u(x), \quad \text{for all } x \right\}$$

based on a pair of functions $(\ell(x), u(x))$. In either case, we would like to have

$$\mathbb{P}_f(f \in \mathcal{B}_n) \geq 1 - \alpha \tag{4.25}$$

for all $f \in \mathcal{F}$ where $\mathcal{F}$ is some large class of functions. In practice, it may be difficult to find $\mathcal{B}_n$ so that (4.25) is true exactly. Instead, we may have to settle for some version of (4.25) that holds only approximately. The main point to keep in mind is that an estimate $\widehat{f}_n$ without some sort of confidence set is usually not useful.

4.5   The Curse of Dimensionality

A problem that occurs with smoothing methods is the **curse of dimen-sionality**, a term usually attributed to Bellman (1961). Roughly speaking, this means that estimation gets harder very quickly as the dimension of the observations increases.

There are at least two versions of this curse. The first is the computational curse of dimensionality. This refers to the fact that the computational burden of some methods can increase exponentially with dimension. Our focus here, however, is with the second version, which we call the statistical curse of dimensionality: if the data have dimension d, then we need a sample size n that grows exponentially with d. Indeed, in the next few chapters we will see that the mean squared of any nonparametric estimator of a smooth (twice differentiable) curve will typically have mean squared error of the form

$$\text{MSE} \approx \frac{c}{n^{4/(4+d)}}$$

for some $c > 0$. If we want the MSE to be equal to some small number δ, we can set MSE $= \delta$ and solve for n. We find that

$$n \propto \left(\frac{c}{\delta}\right)^{d/4}$$

which grows exponentially with dimension d.

The reason for this phenomenon is that smoothing involves estimating a function $f(x)$ using data points in a local neighborhood of x. But in a high-dimensional problem, the data are very sparse, so local neighborhoods contain very few points.

Consider an example. Suppose we have n data points uniformly distributed on the interval $[-1, 1]$. How many data points will we find in the interval $[-0.1, 0.1]$? The answer is: about $n/10$ points. Now suppose we have n data points on the 10-dimensional unit cube $[0, 1]^{10} = [0, 1] \times \cdots \times [0, 1]$. How many data points will we find in the cube $[-0.1, 0.1]^{10}$? The answer is about

$$n \times \left(\frac{.2}{2}\right)^{10} = \frac{n}{10,000,000,000}.$$

Thus, n has to be huge to ensure that small neighborhoods have any data in them.

The bottom line is that all the methods we will discuss can, in principle, be used in high-dimensional problems. However, even if we can overcome the computational problems, we are still left with the statistical curse of dimensionality. You may be able to compute an estimator but it will not be accurate. In fact, if you compute a confidence interval around the estimate (as you should) it will be distressingly large. This is not a failure of the method. Rather, the confidence interval correctly indicates the intrinsic difficulty of the problem.

4.6 Bibliographic Remarks

There are a number of good texts on smoothing methods including Silverman (1986), Scott (1992), Simonoff (1996), Ruppert et al. (2003), and Fan and Gijbels (1996). Other references can be found at the end of Chapters 5 and 6. Hall (1987) discusses problems with Kullback–Leibler loss. For an extensive discussion on the curse of dimensionality, see Hastie et al. (2001).

4.7 Exercises

1. Let $X_1, \ldots, X_n$ be iid from a distribution F with density f. The likelihood function for f is

$$\mathcal{L}_n(f) = \prod_{i=1}^{n} f(X_i).$$

   If the model is set $\mathcal{F}$ of all probability density functions, what is the maximum likelihood estimator of f?

2. Prove equation (4.10).

3. Let $X_1, \ldots, X_n$ be an IID sample from a $N(\theta, 1)$ distribution with density $f_\theta(x) = (2\pi)^{-1/2} e^{-(x-\theta)^2/2}$. Consider the density estimator $\widehat{f}(x) = f_{\widehat{\theta}}(x)$ where $\widehat{\theta} = \overline{X}_n$ is the sample mean. Find the risk of $\widehat{f}$.

4. Recall that the Kullback–Leibler distance between two densities f and g is $D(f, g) = \int f(x) \log\bigl(f(x)/g(x)\bigr) dx$. Consider a one-dimensional parametric model $\{f_\theta(x) : \theta \in \mathbb{R}\}$. Establish an approximation relationship between L_2 loss for the parameter θ and Kullback–Leibler loss. Specifically, show that $D(f_\theta, f_\psi) \approx (\theta - \psi)^2 I(\theta)/2$ where θ is the true value, ψ is close to θ and $I(\theta)$ denotes the Fisher information.

5. What is the relationship between L_1 loss, L_2 loss and Kullback–Leibler loss?

6. Repeat the derivation leading up to equation (4.21) but take X to have d dimensions. Replace the small interval $[-h/2, h/2]$ with a small, d-dimensional rectangle. Find the value of h that minimizes the MSE. Find out how large n needs so that MSE is equal to 0.1.

7. Download the datasets from the examples in this chapter from the book website. Write programs to compute histograms and regressograms and try them on these datasets.

5
Nonparametric Regression

In this chapter we will study nonparametric regression, also known as "learning a function" in the jargon of machine learning. We are given n pairs of observations $(x_1, Y_1), \ldots, (x_n, Y_n)$ as in Figures 5.1, 5.2 and 5.3. The **response variable** Y is related to the **covariate** x by the equations

$$Y_i = r(x_i) + \epsilon_i, \qquad \mathbb{E}(\epsilon_i) = 0, \quad i = 1, \ldots, n \qquad (5.1)$$

where r is the **regression function**. The variable x is also called a **feature**. We want to estimate (or "learn") the function r under weak assumptions. The estimator of $r(x)$ is denoted by $\hat{r}_n(x)$. We also refer to $\hat{r}_n(x)$ as a **smoother**. At first, we will make the simplifying assumption that the variance $\mathbb{V}(\epsilon_i) = \sigma^2$ does not depend on x. We will relax this assumption later.

In (5.1), we are treating the covariate values x_i as fixed. We could instead treat these as random, in which case we write the data as $(X_1, Y_1), \ldots, (X_n, Y_n)$ and $r(x)$ is then interpreted as the mean of Y conditional on $X = x$, that is,

$$r(x) = \mathbb{E}(Y|X = x). \qquad (5.2)$$

There is little difference in the two approaches and we shall mostly take the "fixed x" approach except where noted.

5.3 Example (CMB data). Recall the CMB (cosmic microwave background radiation) data from Example 4.4.

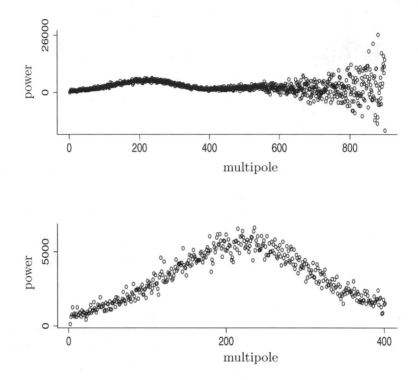

FIGURE 5.1. CMB data. The horizontal axis is the multipole moment, essentially the frequency of fluctuations in the temperature field of the CMB. The vertical axis is the power or strength of the fluctuations at each frequency. The top plot shows the full data set. The bottom plot shows the first 400 data points. The first peak, around $x \approx 200$, is obvious. There may be a second and third peak further to the right.

Figure 5.1 shows the data.[1] The first plot shows 899 data points over the whole range while the second plot shows the first 400 data points. We have noisy measurements Y_i of $r(x_i)$ so the data are of the form (5.1). Our goal is to estimate r. The variance $\mathbb{V}(\epsilon_i)$ is most definitely not constant as a function of x. However, the second plot shows that the constant variance assumption is reasonable for the first 400 data points. It is believed that r may have three peaks over the range of the data. The first peak is obvious from the second plot. The presence of a second or third peak is much less obvious; careful inferences are required to assess the significance of these peaks. ∎

---

[1] If you want to peek ahead, Figure 5.3 shows a nonparametric estimator of the regression function. Note that the vertical scale is different on that plot.

The methods we consider in this chapter are **local regression methods** and **penalization methods**. The former includes **kernel regression** and **local polynomial regression**. The latter leads to methods based on **splines**. In Chapters 8 and 9 we will consider a different approach based on orthogonal functions. All the estimators in this chapter are **linear smoothers**, a point we will discuss in Section 5.2.

Before we plunge into nonparametric regression, we first briefly review ordinary linear regression and its close relative, logistic regression. For more on linear regression, see Weisberg (1985).

5.1   Review of Linear and Logistic Regression

Suppose we have data $(x_1, Y_1), \ldots, (x_n, Y_n)$ where $Y_i \in \mathbb{R}$ and $x_i = (x_{i1}, \ldots, x_{ip})^T \in \mathbb{R}^p$. The **linear regression model** assumes that

$$Y_i = r(x_i) + \epsilon_i \equiv \sum_{j=1}^{p} \beta_j x_{ij} + \epsilon_i, \quad i = 1, \ldots, n, \tag{5.4}$$

where $\mathbb{E}(\epsilon_i) = 0$ and $\mathbb{V}(\epsilon_i) = \sigma^2$.

Warning! Usually, we want to include an intercept in the model so we will adopt the convention that $x_{i1} = 1$.

The **design matrix** X is the $n \times p$ matrix defined by

$$X = \begin{pmatrix} x_{11} & x_{12} & \cdots & x_{1p} \\ x_{21} & x_{22} & \cdots & x_{2p} \\ \vdots & \vdots & \vdots & \vdots \\ x_{n1} & x_{n2} & \cdots & x_{np} \end{pmatrix}.$$

The set $\mathcal{L}$ of vectors that can be obtained as linear combinations of the columns of X, is called the **column space** of X.

Let $Y = (Y_1, \ldots, Y_n)^T$, $\epsilon = (\epsilon_1, \ldots, \epsilon_n)^T$, and $\beta = (\beta_1, \ldots, \beta_p)^T$. We can then write (5.4) as

$$Y = X\beta + \epsilon. \tag{5.5}$$

The **least squares estimator** $\widehat{\beta} = (\widehat{\beta}_1, \ldots, \widehat{\beta}_p)^T$ is the vector that minimizes the **residual sums of squares**

$$\text{RSS} = (Y - X\beta)^T (Y - X\beta) = \sum_{i=1}^{n} \left(Y_i - \sum_{j=1}^{p} x_{ij}\beta_j \right)^2.$$

Assuming that $X^T X$ is invertible, the least squares estimator is

$$\widehat{\beta} = (X^T X)^{-1} X^T Y. \tag{5.6}$$

The estimate of $r(x)$ at $x = (x_1, \ldots, x_p)^T$ is thus

$$\widehat{r}_n(x) = \sum_{j=1}^{p} \widehat{\beta}_j x_j = x^T \widehat{\beta}.$$

It follows that the **fitted values** $\mathbf{r} = (\widehat{r}_n(x_1), \ldots, \widehat{r}_n(x_n))^T$ can be written as

$$\mathbf{r} = X\widehat{\beta} = LY \tag{5.7}$$

where

$$L = X(X^T X)^{-1} X^T \tag{5.8}$$

is called the **hat matrix**. The elements of the vector $\widehat{\epsilon} = Y - \mathbf{r}$ are called the **residuals**. The hat matrix is **symmetric** $L = L^T$ and **idempotent**, $L^2 = L$. It follows that $\mathbf{r}$ is the projection of Y onto the column space $\mathcal{L}$ of X. It can be shown that the number of parameters p is related to the matrix L by way of the equation

$$p = \text{tr}(L) \tag{5.9}$$

where $\text{tr}(L)$ denotes the trace of the matrix L, that is, the sum of its diagonal elements. In nonparametric regression, the number of parameters will be replaced with the **effective degrees of freedom** which will be defined through an equation like (5.9).

Given any $x = (x_1, \ldots, x_p)^T$, we can write

$$\widehat{r}_n(x) = \ell(x)^T Y = \sum_{i=1}^{n} \ell_i(x) Y_i \tag{5.10}$$

where

$$\ell(x)^T = x^T (X^T X)^{-1} X^T.$$

An unbiased estimator of σ^2 is

$$\widehat{\sigma}^2 = \frac{\sum_{i=1}^{n} (Y_i - \widehat{r}_n(x_i))^2}{n - p} = \frac{||\widehat{\epsilon}||^2}{n - p}. \tag{5.11}$$

Next, we construct a confidence band for $r(x)$. We want to find a pair of functions $a(x), b(x)$ such that

$$\mathbb{P}\big(a(x) \leq r(x) \leq b(x) \text{ for all } x\big) \geq 1 - \alpha. \tag{5.12}$$

Since $\widehat{r}_n(x) = \sum_{i=1}^n \ell_i(x)Y_i$, it follows that

$$V(\widehat{r}_n(x)) = \sigma^2 \sum_{i=1}^n \ell_i^2(x) = \sigma^2 ||\ell(x)||^2$$

which suggests using bands of the form

$$I(x) = (a(x), b(x)) \equiv \left(\widehat{r}_n(x) - c\,\widehat{\sigma}||\ell(x)||,\ r_n(x) + c\,\widehat{\sigma}||\ell(x)|| \right) \tag{5.13}$$

for some constant c. The following theorem can be found in Scheffé (1959). Let $F_{p,n-p}$ denote a random variable that has an F distribution with p and $n - p$ degrees of freedom. Let $F_{\alpha;p,n-p}$ denote the upper α quantile for this random variable, i.e., $\mathbb{P}(F_{p,n-p} > F_{\alpha;p,n-p}) = \alpha$.

5.14 Theorem. *The confidence band defined in (5.13) with $c = \sqrt{pF_{\alpha;p,n-p}}$ satisfies (5.12).*

When the Y_i's are not continuous, ordinary linear regression may not be appropriate. For example, suppose that $Y_i \in \{0, 1\}$. In this case, a commonly used parametric model is the **logistic regression model** given by

$$p_i \equiv p_i(\beta) = \mathbb{P}(Y_i = 1) = \frac{e^{\sum_j \beta_j x_{ij}}}{1 + e^{\sum_j \beta_j x_{ij}}}. \tag{5.15}$$

As before, we can include an intercept by requiring that $x_{i1} = 1$ for all i. This model asserts that Y_i is a Bernoulli random variable with mean p_i. The parameters $\beta = (\beta_1, \ldots, \beta_p)^T$ are usually estimated by maximum likelihood. Recall that if $Y \sim \text{Bernoulli}(p)$ then its probability function is $\mathbb{P}(Y = y) \equiv f(y) = p^y(1 - p)^{1-y}$. Thus the likelihood function for the model (5.15) is

$$\mathcal{L}(\beta) = \prod_{i=1}^n p_i^{Y_i}(1 - p_i)^{1-Y_i}. \tag{5.16}$$

The maximum likelihood estimator $\widehat{\beta} = (\widehat{\beta}_1, \ldots, \widehat{\beta}_p)^T$ cannot be found in closed form. However, there is an iterative algorithm called **reweighted least squares** which works as follows.

Reweighted Least Squares Algorithm

Choose starting values $\widehat{\beta} = (\widehat{\beta}_1, \ldots, \widehat{\beta}_p)^T$ and compute p_i using equation (5.15), for $i = 1, \ldots, n$ with β_j replaced by its current estimate $\widehat{\beta}_j$. Iterate the following steps until convergence.

1. Set

$$Z_i = \log\left(\frac{p_i}{1 - p_i} \right) + \frac{Y_i - p_i}{p_i(1 - p_i)}, \quad i = 1, \ldots, n.$$

2. Set the new estimate of β to be

$$\widehat{\beta} = (X^T W X)^{-1} X^T W Z$$

where W is the diagonal matrix with (i, i) element equal to $p_i(1 - p_i)$. This corresponds to doing a (weighted) linear regression of Z on X.

3. Compute the p_is using (5.15) with the current estimate of $\widehat{\beta}$.

Logistic regression and linear regression are special cases of a class of models called **generalized linear models**. See McCullagh and Nelder (1999) for details.

5.2   Linear Smoothers

As we remarked earlier, all the nonparametric estimators in this chapter are linear smoothers. The formal definition is as follows.

5.17 Definition. *An estimator $\widehat{r}_n$ of r is a **linear smoother** if, for each x, there exists a vector $\ell(x) = (\ell_1(x), \ldots, \ell_n(x))^T$ such that*

$$\widehat{r}_n(x) = \sum_{i=1}^{n} \ell_i(x) Y_i. \tag{5.18}$$

Define the vector of **fitted values**

$$\mathbf{r} = (\widehat{r}_n(x_1), \ldots, \widehat{r}_n(x_n))^T \tag{5.19}$$

where $Y = (Y_1, \ldots, Y_n)^T$. It then follows that

$$\mathbf{r} = LY \tag{5.20}$$

where L is an $n \times n$ matrix whose i^{th} row is $\ell(x_i)^T$; thus, $L_{ij} = \ell_j(x_i)$. The entries of the ith row show the weights given to each Y_i in forming the estimate $\widehat{r}_n(x_i)$.

5.21 Definition. *The matrix L is called the **smoothing matrix** or the **hat matrix**. The ith row of L is called the **effective kernel** for estimating $r(x_i)$. In analogy to (5.9), we define the **effective degrees of***

freedom *by*

$$\nu = \text{tr}(L). \tag{5.22}$$

Warning! The reader should not confuse linear smoothers—smoothers of the form (5.18)—with linear regression, in which one assumes that the regression function $r(x)$ is linear.

5.23 Remark. The weights in all the smoothers we will use have the property that, for all x, $\sum_{i=1}^{n} \ell_i(x) = 1$. This implies that the smoother preserves constant curves. Thus, if $Y_i = c$ for all i, then $\hat{r}_n(x) = c$.

5.24 Example (Regressogram). Suppose that $a \le x_i \le b\ i = 1, \ldots, n$. Divide (a, b) into m equally spaced bins denoted by $B_1, B_2, \ldots, B_m$. Define $\hat{r}_n(x)$ by

$$\hat{r}_n(x) = \frac{1}{k_j} \sum_{i: x_i \in B_j} Y_i, \quad \text{for } x \in B_j \tag{5.25}$$

where k_j is the number of points in B_j. In other words, the estimate $\hat{r}_n$ is a step function obtained by averaging the Y_is over each bin. This estimate is called the **regressogram**. An example is given in Figure 4.6. For $x \in B_j$ define $\ell_i(x) = 1/k_j$ if $x_i \in B_j$ and $\ell_i(x) = 0$ otherwise. Thus, $\hat{r}_n(x) = \sum_{i=1}^{n} Y_i \ell_i(x)$. The vector of weights $\ell(x)$ looks like this:

$$\ell(x)^T = \left(0, 0, \ldots, 0, \frac{1}{k_j}, \ldots, \frac{1}{k_j}, 0, \ldots, 0 \right).$$

To see what the smoothing matrix L looks like, suppose that $n = 9$, $m = 3$ and $k_1 = k_2 = k_3 = 3$. Then,

$$L = \frac{1}{3} \times \begin{pmatrix} 1 & 1 & 1 & 0 & 0 & 0 & 0 & 0 & 0 \\ 1 & 1 & 1 & 0 & 0 & 0 & 0 & 0 & 0 \\ 1 & 1 & 1 & 0 & 0 & 0 & 0 & 0 & 0 \\ 0 & 0 & 0 & 1 & 1 & 1 & 0 & 0 & 0 \\ 0 & 0 & 0 & 1 & 1 & 1 & 0 & 0 & 0 \\ 0 & 0 & 0 & 1 & 1 & 1 & 0 & 0 & 0 \\ 0 & 0 & 0 & 0 & 0 & 0 & 1 & 1 & 1 \\ 0 & 0 & 0 & 0 & 0 & 0 & 1 & 1 & 1 \\ 0 & 0 & 0 & 0 & 0 & 0 & 1 & 1 & 1 \end{pmatrix}.$$

In general, it is easy to see that there are $\nu = \text{tr}(L) = m$ effective degrees of freedom. The binwidth $h = (b - a)/m$ controls how smooth the estimate is. and the smoothing matrix L has the form ∎

5.26 Example (Local averages). Fix $h > 0$ and let $B_x = \{i : |x_i - x| \leq h\}$. Let n_x be the number of points in B_x. For any x for which $n_x > 0$ define

$$\widehat{r}_n(x) = \frac{1}{n_x} \sum_{i \in B_x} Y_i.$$

This is the **local average estimator** of $r(x)$, a special case of the kernel estimator discussed shortly. In this case, $\widehat{r}_n(x) = \sum_{i=1}^{n} Y_i \ell_i(x)$ where $\ell_i(x) = 1/n_x$ if $|x_i - x| \leq h$ and $\ell_i(x) = 0$ otherwise. As a simple example, suppose that $n = 9$, $x_i = i/9$ and $h = 1/9$. Then,

$$L = \begin{pmatrix} 1/2 & 1/2 & 0 & 0 & 0 & 0 & 0 & 0 & 0 \\ 1/3 & 1/3 & 1/3 & 0 & 0 & 0 & 0 & 0 & 0 \\ 0 & 1/3 & 1/3 & 1/3 & 0 & 0 & 0 & 0 & 0 \\ 0 & 0 & 1/3 & 1/3 & 1/3 & 0 & 0 & 0 & 0 \\ 0 & 0 & 0 & 1/3 & 1/3 & 1/3 & 0 & 0 & 0 \\ 0 & 0 & 0 & 0 & 1/3 & 1/3 & 1/3 & 0 & 0 \\ 0 & 0 & 0 & 0 & 0 & 1/3 & 1/3 & 1/3 & 0 \\ 0 & 0 & 0 & 0 & 0 & 0 & 1/3 & 1/3 & 1/3 \\ 0 & 0 & 0 & 0 & 0 & 0 & 0 & 1/2 & 1/2 \end{pmatrix}. \blacksquare$$

5.3   Choosing the Smoothing Parameter

The smoothers we will use will depend on some smoothing parameter h and we will need some way of choosing h. As in Chapter 4, define the risk (mean squared error)

$$R(h) = \mathbb{E}\left(\frac{1}{n} \sum_{i=1}^{n} (\widehat{r}_n(x_i) - r(x_i))^2\right). \tag{5.27}$$

Ideally, we would like to choose h to minimize $R(h)$ but $R(h)$ depends on the unknown function $r(x)$. Instead, we will minimize an estimate $\widehat{R}(h)$ of $R(h)$. As a first guess, we might use the average residual sums of squares, also called the **training error**

$$\frac{1}{n} \sum_{i=1}^{n} (Y_i - \widehat{r}_n(x_i))^2 \tag{5.28}$$

to estimate $R(h)$. This turns out to be a poor estimate of $R(h)$: it is biased downwards and typically leads to undersmoothing (overfitting). The reason is that we are using the data twice: to estimate the function and to estimate the risk. The function estimate is chosen to make $\sum_{i=1}^{n} (Y_i - \widehat{r}_n(x_i))^2$ small so this will tend to underestimate the risk.

We will estimate the risk using the leave-one-out cross-validation score which is defined as follows.

5.29 Definition. *The* **leave-one-out cross-validation score** *is defined by*

$$\text{CV} = \widehat{R}(h) = \frac{1}{n}\sum_{i=1}^{n}(Y_i - \widehat{r}_{(-i)}(x_i))^2 \tag{5.30}$$

where $\widehat{r}_{(-i)}$ is the estimator obtained by omitting the ith pair (x_i, Y_i).

As stated above, Definition 5.29 is incomplete. We have not said what we mean precisely by $\widehat{r}_{(-i)}$. We shall define

$$\widehat{r}_{(-i)}(x) = \sum_{j=1}^{n} Y_j \ell_{j,(-i)}(x) \tag{5.31}$$

where

$$\ell_{j,(-i)}(x) = \begin{cases} 0 & \text{if } j = i \\ \frac{\ell_j(x)}{\sum_{k \neq i} \ell_k(x)} & \text{if } j \neq i. \end{cases} \tag{5.32}$$

In other words we set the weight on x_i to 0 and renormalize the other weights to sum to one. For all the methods in this chapter (kernel regression, local polynomials, smoothing splines) this form for $\widehat{r}_{(-i)}$ can actually be derived as a property of the method, rather than a matter of definition. But it is simpler to treat this as a definition.

The intuition for cross-validation is as follows. Note that

$$\begin{aligned} \mathbb{E}(Y_i - \widehat{r}_{(-i)}(x_i))^2 &= \mathbb{E}(Y_i - r(x_i) + r(x_i) - \widehat{r}_{(-i)}(x_i))^2 \\ &= \sigma^2 + \mathbb{E}(r(x_i) - \widehat{r}_{(-i)}(x_i))^2 \\ &\approx \sigma^2 + \mathbb{E}(r(x_i) - \widehat{r}_n(x_i))^2 \end{aligned}$$

and hence, recalling (4.15),

$$\mathbb{E}(\widehat{R}) \approx R + \sigma^2 = \text{predictive risk.} \tag{5.33}$$

Thus the cross-validation score is a nearly unbiased estimate of the risk.

It seems that it might be time consuming to evaluate $\widehat{R}(h)$ since we apparently need to recompute the estimator after dropping out each observation. Fortunately, there is a shortcut formula for computing $\widehat{R}$ for linear smoothers.

5.34 Theorem. *Let $\widehat{r}_n$ be a linear smoother. Then the leave-one-out cross-validation score $\widehat{R}(h)$ can be written as*

$$\widehat{R}(h) = \frac{1}{n} \sum_{i=1}^{n} \left(\frac{Y_i - \widehat{r}_n(x_i)}{1 - L_{ii}} \right)^2 \tag{5.35}$$

where $L_{ii} = \ell_i(x_i)$ is the ith diagonal element of the smoothing matrix L.

The smoothing parameter h can then be chosen by minimizing $\widehat{R}(h)$.

Warning! You can't assume that $\widehat{R}(h)$ always has a well-defined minimum. You should always plot $\widehat{R}(h)$ as a function of h.

Rather than minimizing the cross-validation score, an alternative is to use an approximation called **generalized cross-validation**[2] in which each L_{ii} in equation (5.35) is replaced with its average $n^{-1} \sum_{i=1}^{n} L_{ii} = \nu/n$ where $\nu = \text{tr}(L)$ is the effective degrees of freedom. Thus, we would minimize

$$\text{GCV}(h) = \frac{1}{n} \sum_{i=1}^{n} \left(\frac{Y_i - \widehat{r}_n(x_i)}{1 - \nu/n} \right)^2. \tag{5.36}$$

Usually, the bandwidth that minimizes the generalized cross-validation score is close to the bandwidth that minimizes the cross-validation score.

Using the approximation $(1-x)^{-2} \approx 1 + 2x$ we see that

$$\text{GCV}(h) \approx \frac{1}{n} \sum_{i=1}^{n} (Y_i - \widehat{r}_n(x_i))^2 + \frac{2\nu\widehat{\sigma}^2}{n} \equiv C_p \tag{5.37}$$

where $\widehat{\sigma}^2 = n^{-1} \sum_{i=1}^{n} (Y_i - \widehat{r}_n(x_i))^2$. Equation (5.37) is known as the C_p **statistic**[3] which was originally proposed by Colin Mallows as a criterion for variable selection in linear regression. More generally, many common bandwidth selection criteria can be written in the form

$$B(h) = \Xi(n, h) \times \frac{1}{n} \sum_{i=1}^{n} (Y_i - \widehat{r}_n(x_i))^2 \tag{5.38}$$

for different choices of $\Xi(n, h)$. See Härdle et al. (1988) for details. Moreover, Härdle et al. (1988) prove, under appropriate conditions, the following facts about the minimizer $\widehat{h}$ of $B(h)$. Let $\widehat{h}_0$ minimize the loss $L(\widehat{h}) = n^{-1} \sum_{i=1}^{n} (\widehat{r}_n(x_i) - r(x_i))^2$, and let h_0 minimize the risk. Then all of $\widehat{h}, \widehat{h}_0$, and

---

[2]Generalized cross-validation has certain invariance properties not shared by leave-one-out cross-validation. In practice, however, the two are usually similar.

[3]Actually, this is not quite the C_p formula. Usually, the estimate (5.86) for σ^2 is used.

h_0 tend to 0 at rate $n^{-1/5}$. Also, for certain positive constants $C_1, C_2, \sigma_1, \sigma_2$,

$$n^{3/10}(\widehat{h} - \widehat{h}_0) \rightsquigarrow N(0, \sigma_1^2), \quad n(L(\widehat{h}) - L(\widehat{h}_0)) \rightsquigarrow C_1 \chi_1^2$$
$$n^{3/10}(\widehat{h}_0 - h_0) \rightsquigarrow N(0, \sigma_2^2), \quad n(L(h_0) - L(\widehat{h}_0)) \rightsquigarrow C_2 \chi_1^2.$$

Thus, the relative rate of convergence of $\widehat{h}$ is

$$\frac{\widehat{h} - \widehat{h}_0}{\widehat{h}_0} = O_P\left(\frac{n^{3/10}}{n^{1/5}}\right) = O_P(n^{-1/10}).$$

This slow rate shows that estimating the bandwidth is difficult. This rate is intrinsic to the problem of bandwidth selection since it is also true that

$$\frac{\widehat{h}_0 - h_0}{h_0} = O_P\left(\frac{n^{3/10}}{n^{1/5}}\right) = O_P(n^{-1/10}).$$

5.4   Local Regression

Now we turn to local nonparametric regression. Suppose that $x_i \in \mathbb{R}$ is scalar and consider the regression model (5.1). In this section we consider estimators of $r(x)$ obtained by taking a weighted average of the Y_is, giving higher weight to those points near x. We begin with the kernel regression estimator.

5.39 Definition. Let $h > 0$ be a positive number, called the **bandwidth**. The **Nadaraya–Watson kernel estimator** is defined by

$$\widehat{r}_n(x) = \sum_{i=1}^{n} \ell_i(x) Y_i \tag{5.40}$$

where K is a kernel (as defined in Section 4.2) and the weights $\ell_i(x)$ are given by

$$\ell_i(x) = \frac{K\left(\frac{x - x_i}{h}\right)}{\sum_{j=1}^{n} K\left(\frac{x - x_j}{h}\right)}. \tag{5.41}$$

5.42 Remark. The local average estimator in Example 5.26 is a kernel estimator based on the boxcar kernel.

5.43 Example (CMB data). Recall the CMB data from Figure 5.1. Figure 5.2 shows four different kernel regression fits (using just the first 400 data points) based on increasing bandwidths. The top two plots are based on small bandwidths and the fits are too rough. The bottom right plot is based on large bandwidth and the fit is too smooth. The bottom left plot is just right. The bottom right plot also shows the presence of bias near the boundaries. As we shall see, this is a general feature of kernel regression. The bottom plot in Figure 5.3 shows a kernel fit to all the data points. The bandwidth was chosen by cross-validation. ■

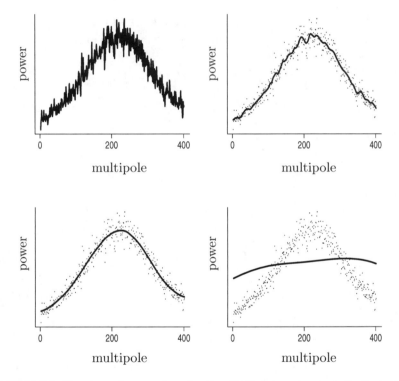

FIGURE 5.2. Four kernel regressions for the CMB data using just the first 400 data points. The bandwidths used were $h = 1$ (top left), $h = 10$ (top right), $h = 50$ (bottom left), $h = 200$ (bottom right). As the bandwidth h increases, the estimated function goes from being too rough to too smooth.

The choice of kernel K is not too important. Estimates obtained by using different kernels are usually numerically very similar. This observation is confirmed by theoretical calculations which show that the risk is very insensitive to the choice of kernel; see Section 6.2.3 of Scott (1992). We will often use the tricube kernel in examples. What does matter much more is the choice of bandwidth h which controls the amount of smoothing. Small bandwidths give very rough estimates while larger bandwidths give smoother estimates. In general, we will let the bandwidth depend on the sample size so we sometimes write h_n.

The following theorem shows how the bandwidth affects the estimator. To state these results we need to make some assumption about the behavior of $x_1, \ldots, x_n$ as n increases. For the purposes of the theorem, we will assume that these are random draws from some density f.

5.44 Theorem. *The risk (using integrated squared error loss) of the Nadaraya–Watson kernel estimator is*

$$R(\widehat{r}_n, r) = \frac{h_n^4}{4} \left(\int x^2 K(x) dx \right)^2 \int \left(r''(x) + 2r'(x) \frac{f'(x)}{f(x)} \right)^2 dx$$

$$+ \frac{\sigma^2 \int K^2(x) dx}{n h_n} \int \frac{1}{f(x)} dx + o(n h_n^{-1}) + o(h_n^4) \quad (5.45)$$

as $h_n \to 0$ and $n h_n \to \infty$.

The first term in (5.45) is the squared bias and the second term is the variance. What is especially notable is the presence of the term

$$2r'(x) \frac{f'(x)}{f(x)} \quad (5.46)$$

in the bias. We call (5.46) the **design bias** since it depends on the design, that is, the distribution of the x_i's. This means that the bias is sensitive to the position of the x_is. Furthermore, it can be shown that kernel estimators also have high bias near the boundaries. This is known as **boundary bias**. We will see that we can reduce these biases by using a refinement called local polynomial regression.

If we differentiate (5.45) and set the result equal to 0, we find that the optimal bandwidth h_* is

$$h_* = \left(\frac{1}{n} \right)^{1/5} \left(\frac{\sigma^2 \int K^2(x) dx \int dx/f(x)}{(\int x^2 K^2(x) dx)^2 \int \left(r''(x) + 2r'(x) \frac{f'(x)}{f(x)} \right)^2 dx} \right)^{1/5} . \quad (5.47)$$

Thus, $h_* = O(n^{-1/5})$. Plugging h_* back into (5.45) we see that the risk decreases at rate $O(n^{-4/5})$. In (most) parametric models, the risk of the maximum likelihood estimator decreases to 0 at rate $1/n$. The slower rate $n^{-4/5}$ is the price of using nonparametric methods. In practice, we cannot use the bandwidth given in (5.47) since h_* depends on the unknown function r. Instead, we use leave-one-out cross-validation as described in Theorem 5.34.

5.48 Example. Figure 5.3 shows the cross-validation score for the CMB example as a function of the effective degrees of freedom. The optimal smoothing parameter was chosen to minimize this score. The resulting fit is also shown in the figure. Note that the fit gets quite variable to the right. Later we will deal with the nonconstant variance and we will add confidence bands to the fit. ∎

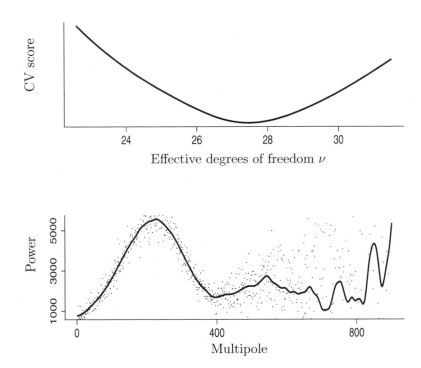

FIGURE 5.3. Top: The cross-validation (CV) score as a function of the effective degrees of freedom. Bottom: the kernel fit using the bandwidth that minimizes the cross-validation score.

LOCAL POLYNOMIALS. Kernel estimators suffer from boundary bias and design bias. These problems can be alleviated by using a generalization of kernel regression called **local polynomial regression**.

To motivate this estimator, first consider choosing an estimator $a \equiv \widehat{r}_n(x)$ to minimize the sums of squares $\sum_{i=1}^{n}(Y_i - a)^2$. The solution is the constant function $\widehat{r}_n(x) = \overline{Y}$ which is obviously not a good estimator of $r(x)$. Now define the weight function $w_i(x) = K((x_i - x)/h)$ and choose $a \equiv \widehat{r}_n(x)$ to minimize the **weighted sums of squares**

$$\sum_{i=1}^{n} w_i(x)(Y_i - a)^2. \tag{5.49}$$

From elementary calculus, we see that the solution is

$$\widehat{r}_n(x) \equiv \frac{\sum_{i=1}^{n} w_i(x)Y_i}{\sum_{i=1}^{n} w_i(x)}$$

which is exactly the kernel regression estimator. This gives us an interesting interpretation of the kernel estimator: it is a locally constant estimator, obtained from locally weighted least squares.

This suggests that we might improve the estimator by using a **local polynomial** of degree p instead of a local constant. Let x be some fixed value at which we want to estimate $r(x)$. For values u in a neighborhood of x, define the polynomial

$$P_x(u; a) = a_0 + a_1(u - x) + \frac{a_2}{2!}(u - x)^2 + \cdots + \frac{a_p}{p!}(u - x)^p. \tag{5.50}$$

We can approximate a smooth regression function $r(u)$ in a neighborhood of the target value x by the polynomial:

$$r(u) \approx P_x(u; a). \tag{5.51}$$

We estimate $a = (a_0, \ldots, a_p)^T$ by choosing $\widehat{a} = (\widehat{a}_0, \ldots, \widehat{a}_p)^T$ to minimize the locally weighted sums of squares

$$\sum_{i=1}^{n} w_i(x) \left(Y_i - P_x(X_i; a)\right)^2. \tag{5.52}$$

The estimator $\widehat{a}$ depends on the target value x so we write $\widehat{a}(x) = (\widehat{a}_0(x), \ldots , \widehat{a}_p(x))^T$ if we want to make this dependence explicit. The local estimate of r is

$$\widehat{r}_n(u) = P_x(u; \widehat{a}).$$

In particular, at the target value $u = x$ we have

$$\widehat{r}_n(x) = P_x(x; \widehat{a}) = \widehat{a}_0(x). \tag{5.53}$$

Warning! Although $\widehat{r}_n(x)$ only depends on $\widehat{a}_0(x)$, this is not equivalent to simply fitting a local constant.

Setting $p = 0$ gives back the kernel estimator. The special case where $p = 1$ is called **local linear regression** and this is the version we recommend as a default choice. As we shall see, local polynomial estimators, and in particular local linear estimators, have some remarkable properties as shown by Fan (1992) and Hastie and Loader (1993). Many of the results that follow are from those papers.

To find $\widehat{a}(x)$, it is helpful to re-express the problem in vector notation. Let

$$X_x = \begin{pmatrix} 1 & x_1 - x & \cdots & \frac{(x_1-x)^p}{p!} \\ 1 & x_2 - x & \cdots & \frac{(x_2-x)^p}{p!} \\ \vdots & \vdots & \ddots & \vdots \\ 1 & x_n - x & \cdots & \frac{(x_n-x)^p}{p!} \end{pmatrix} \tag{5.54}$$

and let W_x be the $n \times n$ diagonal matrix whose (i, i) component is $w_i(x)$. We can rewrite (5.52) as

$$(Y - X_x a)^T W_x (Y - X_x a). \tag{5.55}$$

Minimizing (5.55) gives the weighted least squares estimator

$$\widehat{a}(x) = (X_x^T W_x X_x)^{-1} X_x^T W_x Y. \tag{5.56}$$

In particular, $\widehat{r}_n(x) = \widehat{a}_0(x)$ is the inner product of the first row of $(X_x^T W_x X_x)^{-1} X_x^T W_x$ with Y. Thus we have:

5.57 Theorem. *The local polynomial regression estimate is*

$$\widehat{r}_n(x) = \sum_{i=1}^{n} \ell_i(x) Y_i \tag{5.58}$$

where $\ell(x)^T = (\ell_1(x), \ldots, \ell_n(x))$,

$$\ell(x)^T = e_1^T (X_x^T W_x X_x)^{-1} X_x^T W_x,$$

$e_1 = (1, 0, \ldots, 0)^T$ *and* X_x *and* W_x *are defined in (5.54). This estimator has mean*

$$\mathbb{E}(\widehat{r}_n(x)) = \sum_{i=1}^{n} \ell_i(x) r(x_i)$$

and variance

$$\mathbb{V}(\widehat{r}_n(x)) = \sigma^2 \sum_{i=1}^{n} \ell_i(x)^2 = \sigma^2 \|\ell(x)\|^2.$$

Once again, our estimate is a linear smoother and we can choose the bandwidth by minimizing the cross-validation formula given in Theorem 5.34.

5.59 Example (LIDAR). These data were introduced in Example 4.5. Figure 5.4 shows the 221 observations. The top left plot shows the data and the fitted function using local linear regression. The cross-validation curve (not shown) has a well-defined minimum at $h \approx 37$ corresponding to 9 effective degrees of freedom. The fitted function uses this bandwidth. The top right plot shows the residuals. There is clear heteroscedasticity (nonconstant variance). The bottom left plot shows the estimate of $\sigma(x)$ using the method in Section 5.6. (with $h = 146$ chosen via cross-validation). Next we use the method in Section 5.7 to compute 95 percent confidence bands. The resulting bands are shown in the lower right plot. As expected, there is much greater uncertainty for larger values of the covariate. ■

Local Linear Smoothing

5.60 Theorem. *When* $p = 1$, $\widehat{r}_n(x) = \sum_{i=1}^n \ell_i(x) Y_i$ *where*

$$\ell_i(x) = \frac{b_i(x)}{\sum_{j=1}^n b_j(x)},$$

$$b_i(x) = K\left(\frac{x_i - x}{h}\right)\left(S_{n,2}(x) - (x_i - x)S_{n,1}(x)\right) \qquad (5.61)$$

and

$$S_{n,j}(x) = \sum_{i=1}^n K\left(\frac{x_i - x}{h}\right)(x_i - x)^j, \quad j = 1, 2.$$

5.62 Example. Figure 5.5 shows the local regression for the CMB data for $p = 0$ and $p = 1$. The bottom plots zoom in on the left boundary. Note that for $p = 0$ (the kernel estimator), the fit is poor near the boundaries due to boundary bias. ■

5.63 Example (Doppler function). Let

$$r(x) = \sqrt{x(1-x)} \sin\left(\frac{2.1\pi}{x + .05}\right), \qquad 0 \le x \le 1 \qquad (5.64)$$

which is called the **Doppler function**. This function is difficult to estimate and provides a good test case for nonparametric regression methods. The function is spatially inhomogeneous which means that its smoothness (second

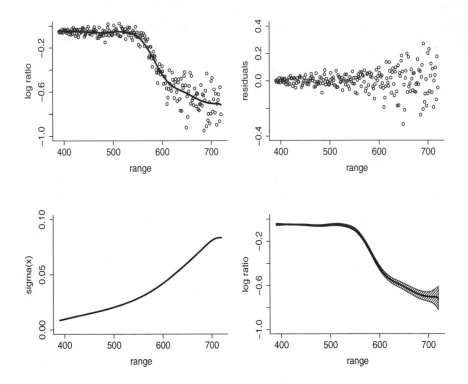

FIGURE 5.4. The LIDAR data from Example 5.59. Top left: data and the fitted function using local linear regression with $h \approx 37$ (chosen by cross-validation). Top right: the residuals. Bottom left: estimate of $\sigma(x)$. Bottom right: 95 percent confidence bands.

derivative) varies over x. The function is plotted in the top left plot of Figure 5.6. The top right plot shows 1000 data points simulated from $Y_i = r(i/n) + \sigma \epsilon_i$ with $\sigma = .1$ and $\epsilon_i \sim N(0, 1)$. The bottom left plot shows the cross-validation score versus the effective degrees of freedom using local linear regression. The minimum occurred at 166 degrees of freedom corresponding to a bandwidth of .005. The fitted function is shown in the bottom right plot. The fit has high effective degrees of freedom and hence the fitted function is very wiggly. This is because the estimate is trying to fit the rapid fluctuations of the function near $x = 0$. If we used more smoothing, the right-hand side of the fit would look better at the cost of missing the structure near $x = 0$. This is always a problem when estimating spatially inhomogeneous functions. We'll discuss that further when we talk about wavelets in Chapter 9. ■

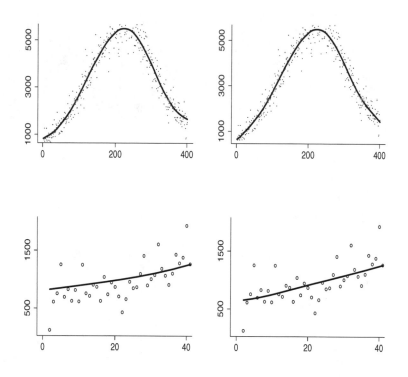

FIGURE 5.5. Locally weighted regressions using local polynomials of order $p = 0$ (top left) and $p = 1$ (top right). The bottom plots show the left boundary in more detail ($p = 0$ bottom left and $p = 1$ bottom right). Notice that the boundary bias is reduced by using local linear estimation ($p = 1$).

The following theorem gives the large sample behavior of the risk of the local linear estimator and shows why local linear regression is better than kernel regression. A proof can be found in Fan (1992) and Fan and Gijbels (1996).

5.65 Theorem. *Let* $Y_i = r(X_i) + \sigma(X_i)\epsilon_i$ *for* $i = 1, \ldots, n$ *and* $a \leq X_i \leq b$. *Assume that* $X_1, \ldots, X_n$ *are a sample from a distribution with density* f *and that (i)* $f(x) > 0$, *(ii)* f, r'' *and* σ^2 *are continuous in a neighborhood of* x, *and (iii)* $h_n \to 0$ *and* $nh_n \to \infty$. *Let* $x \in (a, b)$. *Given* $X_1, \ldots, X_n$, *we have the following: the local linear estimator and the kernel estimator both have variance*

$$\frac{\sigma^2(x)}{f(x)nh_n} \int K^2(u)du + o_P\left(\frac{1}{nh_n}\right). \tag{5.66}$$

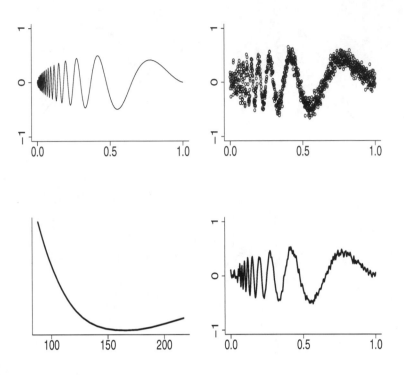

FIGURE 5.6. The Doppler function estimated by local linear regression. The function (top left), the data (top right), the cross-validation score versus effective degrees of freedom (bottom left), and the fitted function (bottom right).

The Nadaraya–Watson kernel estimator has bias

$$h_n^2 \left(\frac{1}{2} r''(x) + \frac{r'(x) f'(x)}{f(x)} \right) \int u^2 K(u) du + o_P(h^2) \tag{5.67}$$

whereas the local linear estimator has asymptotic bias

$$h_n^2 \frac{1}{2} r''(x) \int u^2 K(u) du \; + o_P(h^2) \tag{5.68}$$

Thus, the local linear estimator is free from design bias. At the boundary points a and b, the Nadaraya–Watson kernel estimator has asymptotic bias of order h_n while the local linear estimator has bias of order h_n^2. In this sense, local linear estimation eliminates boundary bias.

5.69 Remark. The above result holds more generally for local polynomials of order p. Generally, taking p odd reduces design bias and boundary bias without increasing variance.

5.5   Penalized Regression, Regularization and Splines

Consider once again the regression model

$$Y_i = r(x_i) + \epsilon_i$$

and suppose we estimate r by choosing $\widehat{r}_n(x)$ to minimize the sums of squares

$$\sum_{i=1}^{n}(Y_i - \widehat{r}_n(x_i))^2.$$

Minimizing over all linear functions (i.e., functions of the form $\beta_0 + \beta_1 x$) yields the least squares estimator. Minimizing over all functions yields a function that interpolates the data. In the previous section we avoided these two extreme solutions by replacing the sums of squares with a locally weighted sums of squares. An alternative way to get solutions in between these extremes is to minimize the **penalized sums of squares**

$$M(\lambda) = \sum_{i}(Y_i - \widehat{r}_n(x_i))^2 + \lambda J(r) \qquad (5.70)$$

where $J(r)$ is some **roughness penalty**. Adding a penalty term to the criterion we are optimizing is sometimes called **regularization**.

We will focus on the special case

$$J(r) = \int (r''(x))^2 dx. \qquad (5.71)$$

The parameter λ controls the trade-off between fit (the first term of 5.70) and the penalty. Let $\widehat{r}_n$ denote the function that minimizes $M(\lambda)$. When $\lambda = 0$, the solution is the interpolating function. When $\lambda \to \infty$, $\widehat{r}_n$ converges to the least squares line. The parameter λ controls the amount of smoothing. What does $\widehat{r}_n$ look like for $0 < \lambda < \infty$? To answer this question, we need to define splines.

A spline is a special piecewise polynomial.[4] The most commonly used splines are piecewise cubic splines.

5.72 Definition. *Let $\xi_1 < \xi_2 < \cdots < \xi_k$ be a set of ordered points—called* **knots**—*contained in some interval (a, b). A* **cubic spline** *is a continuous function r such that (i) r is a cubic polynomial over (ξ_1, ξ_2), $\ldots$ and (ii) r has continuous first and second derivatives at the knots. More generally,*

---

[4]More details on splines can be found in Wahba (1990).

> an Mth-**order spline** *is a piecewise $M - 1$ degree polynomial with $M - 2$ continuous derivatives at the knots. A spline that is linear beyond the boundary knots is called a* **natural spline**.

Cubic splines ($M = 4$) are the most common splines used in practice. They arise naturally in the penalized regression framework as the following theorem shows.

5.73 Theorem. *The function $\widehat{r}_n(x)$ that minimizes $M(\lambda)$ with penalty (5.71) is a natural cubic spline with knots at the data points. The estimator $\widehat{r}_n$ is called a* **smoothing spline**.

The theorem above does not give an explicit form for $\widehat{r}_n$. To do so, we will construct a basis for the set of splines.

5.74 Theorem. *Let $\xi_1 < \xi_2 < \cdots < \xi_k$ be knots contained in an interval (a, b). Define $h_1(x) = 1$, $h_2(x) = x$, $h_3(x) = x^2$, $h_4(x) = x^3$, $h_j(x) = (x - \xi_{j-4})_+^3$ for $j = 5, \ldots, k + 4$. The functions $\{h_1, \ldots, h_{k+4}\}$ form a basis for the set of cubic splines at these knots, called the* **truncated power basis**. *Thus, any cubic spline $r(x)$ with these knots can be written as*

$$r(x) = \sum_{j=1}^{k+4} \beta_j h_j(x). \tag{5.75}$$

We now introduce a different basis for the set of natural splines called the **B-spline basis** that is particularly well suited for computation. These are defined as follows.

Let $\xi_0 = a$ and $\xi_{k+1} = b$. Define new knots $\tau_1, \ldots, \tau_M$ such that

$$\tau_1 \leq \tau_2 \leq \tau_3 \leq \cdots \leq \tau_M \leq \xi_0,$$

$\tau_{j+M} = \xi_j$ for $j = 1, \ldots, k$, and

$$\xi_{k+1} \leq \tau_{k+M+1} \leq \cdots \leq \tau_{k+2M}.$$

The choice of extra knots is arbitrary; usually one takes $\tau_1 = \cdots = \tau_M = \xi_0$ and $\xi_{k+1} = \tau_{k+M+1} = \cdots = \tau_{k+2M}$. We define the basis functions recursively as follows. First we define

$$B_{i,1} = \begin{cases} 1 & \text{if } \tau_i \leq x < \tau_{i+1} \\ 0 & \text{otherwise} \end{cases}$$

for $i = 1, \ldots, k + 2M - 1$. Next, for $m \leq M$ we define

$$B_{i,m} = \frac{x - \tau_i}{\tau_{i+m-1} - \tau_i} B_{i,m-1} + \frac{\tau_{i+m} - x}{\tau_{i+m} - \tau_{i+1}} B_{i+1,m-1}$$

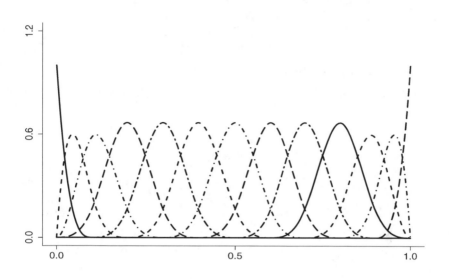

FIGURE 5.7. Cubic B-spline basis using nine equally spaced knots on (0,1).

for $i = 1, \ldots, k + 2M - m$. It is understood that if the denominator is 0, then the function is defined to be 0.

5.76 Theorem. *The functions* $\{B_{i,4}, \ i = 1, \ldots, k + 4\}$ *are a basis for the set of cubic splines. They are called the* **B-spline basis functions**.

The advantage of the B-spline basis functions is that they have compact support which makes it possible to speed up calculations. See Hastie et al. (2001) for details. Figure 5.7 shows the cubic B-spline basis using nine equally spaced knots on (0,1).

We are now in a position to describe the spline estimator in more detail. According to Theorem 5.73, $\widehat{r}$ is a natural cubic spline. Hence, we can write

$$\widehat{r}_n(x) = \sum_{j=1}^{N} \widehat{\beta}_j B_j(x) \tag{5.77}$$

where $B_1, \ldots, B_N$ are a basis for the natural splines (such as the B-splines with $N = n + 4$). Thus, we only need to find the coefficients $\widehat{\beta} = (\widehat{\beta}_1, \ldots, \widehat{\beta}_N)^T$. By expanding r in the basis we can now rewrite the minimization as follows:

$$\text{minimize}: \quad (Y - B\beta)^T (Y - B\beta) + \lambda \beta^T \Omega \beta \tag{5.78}$$

where $B_{ij} = B_j(X_i)$ and $\Omega_{jk} = \int B_j''(x) B_k''(x) dx$.

5.79 Theorem. *The value of β that minimizes (5.78) is*[5]

$$\widehat{\beta} = (B^T B + \lambda \Omega)^{-1} B^T Y. \tag{5.80}$$

Splines are another example of linear smoothers.

5.81 Theorem. *The smoothing spline $\widehat{r}_n(x)$ is a linear smoother, that is, there exist weights $\ell(x)$ such that $\widehat{r}_n(x) = \sum_{i=1}^n Y_i \ell_i(x)$. In particular, the smoothing matrix L is*

$$L = B(B^T B + \lambda \Omega)^{-1} B^T \tag{5.82}$$

and the vector $\mathbf{r}$ of fitted values is given by

$$\mathbf{r} = LY. \tag{5.83}$$

If we had done ordinary linear regression of Y on B, the hat matrix would be $L = B(B^T B)^{-1} B^T$ and the fitted values would interpolate the observed data. The effect of the term $\lambda \Omega$ in (5.82) is to shrink the regression coefficients towards a subspace, which results in a smoother fit. As before, we define the effective degrees of freedom by $\nu = \text{tr}(L)$ and we choose the smoothing parameter λ by minimizing either the cross-validation score (5.35) or the generalized cross-validation score (5.36).

5.84 Example. Figure 5.8 shows the smoothing spline with cross-validation for the CMB data. The effective number of degrees of freedom is 8.8. The fit is smoother than the local regression estimator. This is certainly visually more appealing, but the difference between the two fits is small compared to the width of the confidence bands that we will compute later. ∎

Silverman (1984) proved that spline estimates $\widehat{r}_n(x)$ are approximately kernel estimates in the sense that

$$\ell_i(x) \approx \frac{1}{f(x_i)h(x_i)} K \left(\frac{x_i - x}{h(x_i)} \right)$$

where $f(x)$ is the density of the covariate (treated here as random),

$$h(x) = \left(\frac{\lambda}{nf(x)} \right)^{1/4}$$

and

$$K(t) = \frac{1}{2} \exp \left\{ -\frac{|t|}{\sqrt{2}} \right\} \sin \left(\frac{|t|}{\sqrt{2}} + \frac{\pi}{4} \right).$$

---

[5] If you are familiar with ridge regression then you will recognize this as being similar to ridge regression.

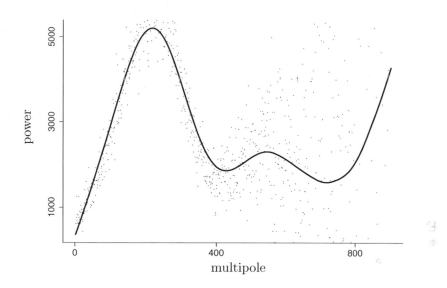

FIGURE 5.8. Smoothing spline for the CMB data. The smoothing parameter was chosen by cross-validation.

Another nonparametric method that uses splines is called **the regression spline method.** Rather than placing a knot at each data point, we instead use fewer knots. We then do ordinary linear regression on the basis matrix B with no regularization. The fitted values for this estimator are $\mathbf{r} = LY$ with $L = B(B^T B)^{-1} B^T$. The difference between this estimate and (5.82) is that the basis matrix B is based on fewer knots and there is no shrinkage factor $\lambda \Omega$. The amount of smoothing is instead controlled by the choice of the number (and placement) of the knots. By using fewer knots, one can save computation time. These spline methods are discussed in detail in Ruppert et al. (2003).

5.6   Variance Estimation

Next we consider several methods for estimating σ^2. For linear smoothers, there is a simple, nearly unbiased estimate of σ^2.

5.85 Theorem. Let $\widehat{r}_n(x)$ be a linear smoother. Let

$$\widehat{\sigma}^2 = \frac{\sum_{i=1}^{n}(Y_i - \widehat{r}(x_i))^2}{n - 2\nu + \widetilde{\nu}} \tag{5.86}$$

where

$$\nu = \text{tr}(L), \quad \tilde{\nu} = \text{tr}(L^T L) = \sum_{i=1}^{n} ||\ell(x_i)||^2.$$

If r is sufficiently smooth, $\nu = o(n)$ and $\tilde{\nu} = o(n)$ then $\hat{\sigma}^2$ is a consistent estimator of σ^2.

We will now outline the proof of this result. Recall that if Y is a random vector and Q is a symmetric matrix then $Y^T Q Y$ is called a **quadratic form** and it is well known that

$$\mathbb{E}(Y^T Q Y) = \text{tr}(QV) + \mu^T Q \mu \tag{5.87}$$

where $V = \mathbb{V}(Y)$ is the covariance matrix of Y and $\mu = \mathbb{E}(Y)$ is the mean vector. Now,

$$Y - \mathbf{r} = Y - LY = (I - L)Y$$

and so

$$\hat{\sigma}^2 = \frac{Y^T \Lambda Y}{\text{tr}(\Lambda)} \tag{5.88}$$

where $\Lambda = (I - L)^T (I - L)$. Hence,

$$\mathbb{E}(\hat{\sigma}^2) = \frac{\mathbb{E}(Y^T \Lambda Y)}{\text{tr}(\Lambda)} = \sigma^2 + \frac{\mathbf{r}^T \Lambda \mathbf{r}}{n - 2\nu + \tilde{\nu}}.$$

Assuming that ν and $\tilde{\nu}$ do not grow too quickly, and that r is smooth, the last term is small for large n and hence $\mathbb{E}(\hat{\sigma}^2) \approx \sigma^2$. Similarly, one can show that $\mathbb{V}(\hat{\sigma}^2) \to 0$.

Here is another estimator, due to Rice (1984). Suppose that the x_is are ordered. Define

$$\hat{\sigma}^2 = \frac{1}{2(n-1)} \sum_{i=1}^{n-1} (Y_{i+1} - Y_i)^2. \tag{5.89}$$

The motivation for this estimator is as follows. Assuming $r(x)$ is smooth, we have $r(x_{i+1}) - r(x_i) \approx 0$ and hence

$$Y_{i+1} - Y_i = \left[r(x_{i+1}) + \epsilon_{i+1} \right] - \left[r(x_i) + \epsilon_i \right] \approx \epsilon_{i+1} - \epsilon_i$$

and hence $(Y_{i+1} - Y_i)^2 \approx \epsilon_{i+1}^2 + \epsilon_i^2 - 2\epsilon_{i+1}\epsilon_i$. Therefore,

$$\begin{aligned} \mathbb{E}(Y_{i+1} - Y_i)^2 &\approx \mathbb{E}(\epsilon_{i+1}^2) + \mathbb{E}(\epsilon_i^2) - 2\mathbb{E}(\epsilon_{i+1})\mathbb{E}(\epsilon_i) \\ &= \mathbb{E}(\epsilon_{i+1}^2) + \mathbb{E}(\epsilon_i^2) = 2\sigma^2. \end{aligned} \tag{5.90}$$

Thus, $\mathbb{E}(\widehat{\sigma}^2) \approx \sigma^2$. A variation of this estimator, due to Gasser et al. (1986) is

$$\widehat{\sigma}^2 = \frac{1}{n-2} \sum_{i=2}^{n-1} c_i^2 \delta_i^2 \qquad (5.91)$$

where

$$\delta_i = a_i Y_{i-1} + b_i Y_{i+1} - Y_i, \qquad a_i = (x_{i+1} - x_i)/(x_{i+1} - x_{i-1}),$$
$$b_i = (x_i - x_{i-1})/(x_{i+1} - x_{i-1}), \qquad c_i^2 = (a_i^2 + b_i^2 + 1)^{-1}.$$

The intuition of this estimator is that it is the average of the residuals that result from fitting a line to the first and third point of each consecutive triple of design points.

5.92 Example. The variance looks roughly constant for the first 400 observations of the CMB data. Using a local linear fit, we applied the two variance estimators. Equation (5.86) yields $\widehat{\sigma}^2 = 408.29$ while equation (5.89) yields $\widehat{\sigma}^2 = 394.55$. ∎

So far we have assumed **homoscedasticity** meaning that $\sigma^2 = \mathbb{V}(\epsilon_i)$ does not vary with x. In the CMB example this is blatantly false. Clearly, σ^2 increases with x so the data are **heteroscedastic**. The function estimate $\widehat{r}_n(x)$ is relatively insensitive to heteroscedasticity. However, when it comes to making confidence bands for $r(x)$, we must take into account the nonconstant variance.

We will take the following approach. See Yu and Jones (2004) and references therein for other approaches. Suppose that

$$Y_i = r(x_i) + \sigma(x_i)\epsilon_i. \qquad (5.93)$$

Let $Z_i = \log(Y_i - r(x_i))^2$ and $\delta_i = \log \epsilon_i^2$. Then,

$$Z_i = \log(\sigma^2(x_i)) + \delta_i. \qquad (5.94)$$

This suggests estimating $\log \sigma^2(x)$ by regressing the log squared residuals on x. We proceed as follows.

Variance Function Estimation

1. Estimate $r(x)$ with any nonparametric method to get an estimate $\widehat{r}_n(x)$.

2. Define $Z_i = \log(Y_i - \widehat{r}_n(x_i))^2$.

3. Regress the Z_i's on the x_i's (again using any nonparametric method)
to get an estimate $\widehat{q}(x)$ of $\log \sigma^2(x)$ and let

$$\widehat{\sigma}^2(x) = e^{\widehat{q}(x)}. \qquad (5.95)$$

5.96 Example. The solid line in Figure 5.9 shows the log of $\widehat{\sigma}^2(x)$ for the CMB
example. I used local linear estimation and I used cross-validation to choose
the bandwidth. The estimated optimal bandwidth for $\widehat{r}_n$ was $h = 42$ while
the estimated optimal bandwidth for the log variance was $h = 160$. In this
example, there turns out to be an independent estimate of $\sigma(x)$. Specifically,
because the physics of the measurement process is well understood, physicists
can compute a reasonably accurate approximation to $\sigma^2(x)$. The log of this
function is the dotted line on the plot. ∎

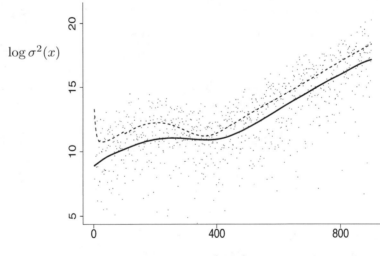

multipole

FIGURE 5.9. The dots are the log squared residuals. The solid line shows the log
of the estimated standard variance $\widehat{\sigma}^2(x)$ as a function of x. The dotted line shows
the log of the true $\sigma^2(x)$ which is known (to reasonable accuracy) through prior
knowledge.

A drawback of this approach is that the log of a very small residual will be
a large outlier. An alternative is to directly smooth the squared residuals. In

this case, one might fit a model of the type discussed in Section 5.10 since the model does not have an additive form and the error will not be Normal.

5.7   Confidence Bands

In this section we will construct confidence bands for $r(x)$. Typically these bands are of the form

$$\widehat{r}_n(x) \pm c \; \mathsf{se}(x) \qquad\qquad (5.97)$$

where $\mathsf{se}(x)$ is an estimate of the standard deviation of $\widehat{r}_n(x)$ and $c > 0$ is some constant. Before we proceed, we discuss a pernicious problem that arises whenever we do smoothing, namely, the bias problem.

THE BIAS PROBLEM. Confidence bands like those in (5.97), are not really confidence bands for $r(x)$, rather, they are confidence bands for $\overline{r}_n(x) = \mathbb{E}(\widehat{r}_n(x))$ which you can think of as a smoothed version of $r(x)$. Getting a confidence set for the true function $r(x)$ is complicated for reasons we now explain.

Denote the mean and standard deviation of $\widehat{r}_n(x)$ by $\overline{r}_n(x)$ and $s_n(x)$. Then,

$$
\begin{aligned}
\frac{\widehat{r}_n(x) - r(x)}{s_n(x)} &= \frac{\widehat{r}_n(x) - \overline{r}_n(x)}{s_n(x)} + \frac{\overline{r}_n(x) - r(x)}{s_n(x)} \\
&= Z_n(x) + \frac{\mathrm{bias}(\widehat{r}_n(x))}{\sqrt{\mathrm{variance}(\widehat{r}_n(x))}}
\end{aligned}
$$

where $Z_n(x) = (\widehat{r}_n(x) - \overline{r}_n(x))/s_n(x)$. Typically, the first term $Z_n(x)$ converges to a standard Normal from which one derives confidence bands. The second term is the bias divided by the standard deviation. In parametric inference, the bias is usually smaller than the standard deviation of the estimator so this term goes to zero as the sample size increases. In nonparametric inference, we have seen that optimal smoothing corresponds to balancing the bias and the standard deviation. The second term does not vanish even with large sample sizes.

> **The presence of this second, nonvanishing term introduces a bias into the Normal limit. The result is that the confidence interval will not be centered around the true function r due to the smoothing bias $\overline{r}_n(x) - r(x)$.**

There are several things we can do about this problem. The first is: live with it. In other words, just accept the fact that the confidence band is for $\overline{r}_n$ not r.

There is nothing wrong with this as long as we are careful when we report the results to make it clear that the inferences are for $\bar{r}_n$ not r. A second approach is to estimate the bias function $\bar{r}_n(x) - r(x)$. This is difficult to do. Indeed, the leading term of the bias is $r''(x)$ and estimating the second derivative of r is much harder than estimating r. This requires introducing extra smoothness conditions which then bring into question the original estimator that did not use this extra smoothness. This has a certain unpleasant circularity to it.[6] A third approach is to **undersmooth.** If we smooth less than the optimal amount then the bias will decrease asymptotically relative to the variance. Unfortunately, there does not seem to be a simple, practical rule for choosing just the right amount of undersmoothing. (See the end of this chapter for more discussion on this point.) We will take the first approach and content ourselves with finding a confidence band for $\bar{r}_n$.

5.98 Example. To understand the implications of estimating $\bar{r}_n$ instead of r, consider the following example. Let

$$r(x) = \phi(x; 2, 1) + \phi(x; 4, 0.5) + \phi(x; 6, 0.1) + \phi(x; 8, 0.05)$$

where $\phi(x; m, s)$ denotes a Normal density function with mean m and variance s^2. Figure 5.10 shows the true function (top left), a locally linear estimate $\hat{r}_n$ (top right) based on 100 observations $Y_i = r(i/10) + .2N(0, 1)$, $i = 1, \ldots, 100$, with bandwidth $h = 0.27$, the function $\bar{r}_n(x) = \mathbb{E}(\hat{r}_n(x))$ (bottom left) and the difference $r(x) - \bar{r}_n(x)$ (bottom right). We see that $\bar{r}_n$ (dashed line) smooths out the peaks. Comparing the top right and bottom left plot, it is clear that $\hat{r}_n(x)$ is actually estimating $\bar{r}_n$ not $r(x)$. Overall, $\bar{r}_n$ is quite similar to $r(x)$ except that $\bar{r}_n$ omits some of the fine details of r. ∎

CONSTRUCTING CONFIDENCE BANDS. Assume that $\hat{r}_n(x)$ is a linear smoother, so that $\hat{r}_n(x) = \sum_{i=1}^n Y_i \ell_i(x)$. Then,

$$\bar{r}(x) = \mathbb{E}(\hat{r}_n(x)) = \sum_{i=1}^n \ell_i(x) r(x_i).$$

For now, assume that $\sigma^2(x) = \sigma^2 = \mathbb{V}(\epsilon_i)$ is constant. Then,

$$\mathbb{V}(\hat{r}_n(x)) = \sigma^2 ||\ell(x)||^2.$$

---

[6] A different approach to estimating bias is discussed in Section 6.4 of Ruppert et al. (2003). However, I am not aware of any theoretical results to justify the resulting confidence bands.

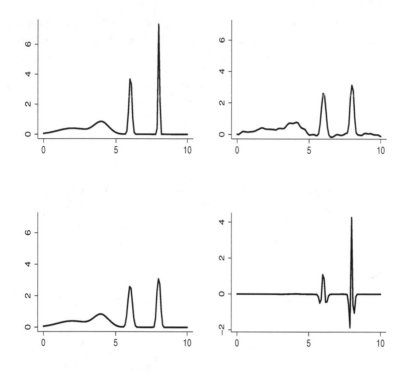

FIGURE 5.10. The true function (top left), an estimate $\widehat{r}_n$ (top right) based on 100 observations, the function $\bar{r}_n(x) = \mathbb{E}(\widehat{r}_n(x))$ (bottom left) and the difference $r(x) - \bar{r}_n(x)$ (bottom right).

We will consider a confidence band for $\bar{r}_n(x)$ of the form

$$I(x) = \left(\widehat{r}_n(x) - c\,\widehat{\sigma}||\ell(x)||,\ \widehat{r}_n(x) + c\,\widehat{\sigma}||\ell(x)||\right) \tag{5.99}$$

for some $c > 0$ and $a \le x \le b$.

We follow the approach in Sun and Loader (1994). First suppose that σ is known. Then,

$$\mathbb{P}\big(\bar{r}(x) \notin I(x) \text{ for some } x \in [a, b]\big) = \mathbb{P}\left(\max_{x\in[a,b]} \frac{|\widehat{r}(x) - \bar{r}(x)|}{\sigma||\ell(x)||} > c\right)$$

$$= \mathbb{P}\left(\max_{x\in[a,b]} \frac{|\sum_i \epsilon_i \ell_i(x)|}{\sigma||\ell(x)||} > c\right) = \mathbb{P}\left(\max_{x\in[a,b]} |W(x)| > c\right)$$

where $W(x) = \sum_{i=1}^{n} Z_i T_i(x)$, $Z_i = \epsilon_i/\sigma \sim N(0,1)$ and $T_i(x) = \ell_i(x)/\|\ell(x)\|$. Now, $W(x)$ is a Gaussian process.[7] To find c, we need to be able to compute the distribution of the maximum of a Gaussian process. Fortunately, this is a well-studied problem. In particular, Sun and Loader (1994) showed that

$$\mathbb{P}\left(\max_x \left|\sum_{i=1}^{n} Z_i T_i(x)\right| > c\right) \approx 2\left(1 - \Phi(c)\right) + \frac{\kappa_0}{\pi} e^{-c^2/2} \qquad (5.100)$$

for large c, where

$$\kappa_0 = \int_a^b \|T'(x)\| dx, \qquad (5.101)$$

$T'(x) = (T_1'(x), \ldots, T_n'(x))$ and $T_i'(x) = \partial T_i(x)/\partial x$. An approximation for κ_0 is given in Exercise 20. Equation (5.100) is called the **tube formula**. A derivation is outlined in the appendix. If we choose c to solve

$$2\left(1 - \Phi(c)\right) + \frac{\kappa_0}{\pi} e^{-c^2/2} = \alpha \qquad (5.102)$$

then we get the desired simultaneous confidence band. If σ is unknown we use an estimate $\hat{\sigma}$. Sun and Loader suggest replacing the right-hand side of (5.100) with

$$\mathbb{P}(|T_m| > c) + \frac{\kappa_0}{\pi}\left(1 + \frac{c^2}{m}\right)^{-m/2}$$

where T_m has a t-distribution with $m = n - \text{tr}(L)$ degrees of freedom. For large n, (5.100) remains an adequate approximation.

Now suppose that $\sigma(x)$ is a function of x. Then,

$$\mathbb{V}(\hat{r}_n(x)) = \sum_{i=1}^{n} \sigma^2(x_i)\ell_i^2(x).$$

In this case we take

$$I(x) = \hat{r}_n(x) \pm c\, s(x) \qquad (5.103)$$

where

$$s(x) = \sqrt{\sum_{i=1}^{n} \hat{\sigma}^2(x_i)\ell_i^2(x)},$$

$\hat{\sigma}(x)$ is an estimate of $\sigma(x)$ and c is the constant defined in (5.102). If $\hat{\sigma}(x)$ varies slowly with x, then $\sigma(x_i) \approx \sigma(x)$ for those i such that $\ell_i(x)$ is large and so

$$s(x) \approx \hat{\sigma}(x)\|\ell(x)\|.$$

---

[7] This means it is a random function such that the vector $(W(x_1), \ldots, W(x_k))$ has a multivariate Normal distribution, for any finite set of points $x_1, \ldots, x_k$.

Thus, an approximate confidence band is

$$I(x) = \widehat{r}_n(x) \pm c\,\widehat{\sigma}(x)\|\ell(x)\|. \tag{5.104}$$

For more details on these methods, see Faraway and Sun (1995).

5.105 Example. Figure 5.11 shows simultaneous 95 percent confidence bands for the CMB data using a local linear fit. The bandwidth was chosen using cross-validation. We find that $\kappa_0 = 38.85$ and $c = 3.33$. In the top plot, we assumed a constant variance when constructing the band. In the bottom plot, we did not assume a constant variance when constructing the band. We see that if we do not take into account the nonconstant variance, we overestimate the uncertainty for small x and we underestimate the uncertainty for large x. ∎

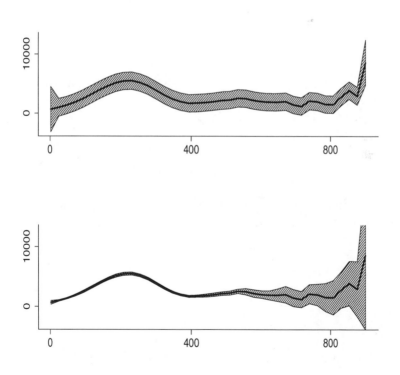

FIGURE 5.11. Local linear fit with simultaneous 95 percent confidence bands. The band in the top plot assumes constant variance σ^2. The band in the bottom plot allows for nonconstant variance $\widehat{\sigma}^2(x)$.

5.106 Remark. We have ignored the uncertainty introduced due to the choice of smoothing parameter. We can adjust for this extra uncertainty as follows. When choosing the smoothing parameter h, restrict the search to a finite set $\mathcal{H}_n$ with $m = m(n)$ points. Construct the confidence band at level α/m. Thus, replace α with α/m on the right-hand side of equation (5.102). Then the Bonferroni inequality[8] guarantees that the coverage is still at least $1 - \alpha$.

5.107 Remark. There is a substantial literature on using bootstrap methods to get confidence bands. This requires a more sophisticated use of the bootstrap than that presented in Chapter 3. See Härdle and Marron (1991), Neumann and Polzehl (1998), Hall (1993), Faraway (1990), and Härdle and Bowman (1988), for example.

5.8   Average Coverage

One might argue that requiring bands to cover the function at all x is too stringent. Wahba (1983), Nychka (1988) and Cummins et al. (2001) introduced a different type of coverage that we refer to as average coverage. Here we discuss a method for constructing average coverage bands based on an idea in Juditsky and Lambert-Lacroix (2003).

Suppose we are estimating $r(x)$ over the interval $[0,1]$. Define the **average coverage** of a band (ℓ, u) by

$$C = \int_0^1 \mathbb{P}\big(r(x) \in [\ell(x), u(x)]\big)\, dx.$$

In Chapters 7 and 8 we present methods for constructing confidence balls for r. These are sets $\mathcal{B}_n(\alpha)$ of the form

$$\mathcal{B}_n(\alpha) = \Big\{r :\ ||\widehat{r}_n - r|| \leq s_n(\alpha)\Big\}$$

such that

$$\mathbb{P}(r \in \mathcal{B}_n(\alpha)) \geq 1 - \alpha.$$

Given such a confidence ball, let

$$\ell(x) = \widehat{r}_n(x) - s_n(\alpha/2)\sqrt{\frac{2}{\alpha}}, \quad u(x) = \widehat{r}_n(x) + s_n(\alpha/2)\sqrt{\frac{2}{\alpha}}. \qquad (5.108)$$

We now show that these bands have average coverage at least $1 - \alpha$. First, note that $C = \mathbb{P}(r(U) \in [\ell(U), u(U)])$ where $U \sim \text{Unif}(0, 1)$ is independent of

---

[8] That is, $\mathbb{P}(A_1 \bigcup \cdots \bigcup A_m) \leq \sum_i \mathbb{P}(A_i)$.

the data. Let A be the event that $r \in \mathcal{B}_n(\alpha/2)$. On the event A, $||\widehat{r}_n - r|| \leq s_n(\alpha/2)$. Writing s_n for $s_n(\alpha/2)$ we have,

$$
\begin{aligned}
1 - C &= \mathbb{P}\left(r(U) \notin [\ell(U), u(U)]\right) = \mathbb{P}\left(|\widehat{r}_n(U) - r(U)| > s_n\sqrt{\frac{2}{\alpha}}\right) \\
&= \mathbb{P}\left(|\widehat{r}_n(U) - r(U)| > s_n\sqrt{\frac{2}{\alpha}}, A\right) + \mathbb{P}\left(|\widehat{r}_n(U) - r(U)| > s_n\sqrt{\frac{2}{\alpha}}, A^c\right) \\
&\leq \mathbb{P}\left(|\widehat{r}_n(U) - r(U)| > s_n\sqrt{\frac{2}{\alpha}}, A\right) + \mathbb{P}(A^c) \\
&\leq \frac{\mathbb{E}I_A|\widehat{r}_n(U) - r(U)|^2}{s_n^2\frac{2}{\alpha}} + \frac{\alpha}{2} = \frac{\mathbb{E}I_A\int_0^1 |\widehat{r}_n(u) - r(u)|^2 du}{s_n^2\frac{2}{\alpha}} + \frac{\alpha}{2} \\
&= \frac{\mathbb{E}I_A||\widehat{r}_n - r||^2}{s_n^2\frac{2}{\alpha}} + \frac{\alpha}{2} \leq \frac{s_n^2}{s_n^2\frac{2}{\alpha}} + \frac{\alpha}{2} \leq \alpha.
\end{aligned}
$$

5.9   Summary of Linear Smoothing

At this point we have covered many points related to linear smoothing methods. It seems like a good time so summarize the steps needed to construct the estimate $\widehat{r}_n$ and a confidence band.

Summary of Linear Smoothing

1. Choose a smoothing method such as local polynomial, spline, etc. This amounts to choosing the form of the weights $\ell(x) = (\ell_1(x), \ldots, \ell_n(x))^T$. A good default choice is local linear smoothing as described in Theorem 5.60.

2. Choose the bandwidth h by cross-validation using (5.35).

3. Estimate the variance function $\widehat{\sigma}^2(x)$ as described in Section 5.6.

4. Find κ_0 from (5.101) and find c from (5.102).

5. An approximate $1 - \alpha$ confidence band for $\overline{r}_n = \mathbb{E}(\widehat{r}_n(x))$ is

$$\widehat{r}_n(x) \pm c\,\widehat{\sigma}(x)\,||\ell(x)||. \qquad (5.109)$$

5.110 Example (LIDAR). Recall the LIDAR data from Example 4.5 and Example 5.59. We find that $\kappa_0 \approx 30$ and $c \approx 3.25$. The resulting bands are shown in the lower right plot. As expected, there is much greater uncertainty for larger values of the covariate. ∎

5.10   Local Likelihood and Exponential Families

If Y is not real valued or ϵ is not Gaussian, then the basic regression model we have been using might not be appropriate. For example, if $Y \in \{0, 1\}$ then it seems natural to use a Bernoulli model. In this section we discuss nonparametric regression for more general models. Before proceeding, we should point out that the basic model often does work well even in cases where Y is not real valued or ϵ is not Gaussian. This is because the asymptotic theory does not really depend on ϵ being Gaussian. Thus, at least for large samples, it is worth considering using the tools we have already developed for these cases.

Recall that Y has an exponential family distribution, given x, if

$$f(y|x) = \exp\left\{ \frac{y\theta(x) - b(\theta(x))}{a(\phi)} + c(y, \phi) \right\} \tag{5.111}$$

for some functions $a(\cdot), b(\cdot)$ and $c(\cdot, \cdot)$. Here, $\theta(\cdot)$ is called the canonical parameter and ϕ is called the dispersion parameter. It then follows that

$$
\begin{aligned}
r(x) &\equiv \mathbb{E}(Y|X = x) = b'(\theta(x)), \\
\sigma^2(x) &\equiv \mathbb{V}(Y|X = x) = a(\phi)b''(\theta(x)).
\end{aligned}
$$

The usual parametric form of this model is

$$g(r(x)) = x^T \beta$$

for some known function g called the **link function**. The model

$$Y|X = x \sim f(y|x), \quad g(\mathbb{E}(Y|X = x)) = x^T \beta \tag{5.112}$$

is called a **generalized linear model**.

For example, if Y given $X = x$ is Binomial $(m, r(x))$ then

$$f(y|x) = \binom{m}{y} r(x)^y (1 - r(x))^{m-y}$$

which has the form (5.111) with

$$\theta(x) = \log \frac{r(x)}{1 - r(x)}, \quad b(\theta) = m \log(1 + e^\theta)$$

and $a(\phi) \equiv 1$. Taking $g(t) = \log(t/(m-t))$ yields the logistic regression model. The parameters β are usually estimated by maximum likelihood.

Let's consider a nonparametric version of logistic regression. For simplicity, we focus on local linear estimation. The data are $(x_1, Y_1), \ldots, (x_n, Y_n)$ where $Y_i \in \{0, 1\}$. We assume that

$$Y_i \sim \text{Bernoulli}(r(x_i))$$

for some smooth function $r(x)$ for which $0 \leq r(x) \leq 1$. Thus, $\mathbb{P}(Y_i = 1 | X_i = x_i) = r(x_i)$ and $\mathbb{P}(Y_i = 0 | X_i = x_i) = 1 - r(x_i)$. The likelihood function is

$$\prod_{i=1}^{n} r(x_i)^{Y_i} (1 - r(x_i))^{1-Y_i}$$

so, with $\xi(x) = \log(r(x)/(1 - r(x)))$, the log-likelihood is

$$\ell(r) = \sum_{i=1}^{n} \ell(Y_i, \xi(x_i)) \tag{5.113}$$

where

$$
\begin{aligned}
\ell(y, \xi) &= \log \left[\left(\frac{e^\xi}{1 + e^\xi} \right)^y \left(\frac{1}{1 + e^\xi} \right)^{1-y} \right] \\
&= y\xi - \log \left(1 + e^\xi \right). \tag{5.114}
\end{aligned}
$$

To estimate the regression function at x we approximate the regression function $r(u)$ for u near x by the local logistic function

$$r(u) \approx \frac{e^{a_0 + a_1(u-x)}}{1 + e^{a_0 + a_1(u-x)}}$$

(compare with (5.15)). Equivalently, we approximate $\log(r(u)/(1-r(u)))$ with $a_0 + a_1(x - u)$. Now define the **local log-likelihood**

$$
\begin{aligned}
\ell_x(a) &= \sum_{i=1}^{n} K \left(\frac{x - X_i}{h} \right) \ell(Y_i, a_0 + a_1(X_i - x)) \\
&= \sum_{i=1}^{n} K \left(\frac{x - X_i}{h} \right) \left(Y_i(a_0 + a_1(X_i - x)) - \log(1 + e^{a_0 + a_1(X_i - x)}) \right).
\end{aligned}
$$

Let $\widehat{a}(x) = (\widehat{a}_0(x), \widehat{a}_1(x))$ maximize ℓ_x which can be found by any convenient optimization routine such as Newton–Raphson. The nonparametric estimate of $r(x)$ is

$$\widehat{r}_n(x) = \frac{e^{\widehat{a}_0(x)}}{1 + e^{\widehat{a}_0(x)}}. \tag{5.115}$$

The bandwidth can be chosen by using the leave-one-out log-likelihood cross-validation

$$\text{CV} = \sum_{i=1}^{n} \ell(Y_i, \widehat{\xi}_{(-i)}(x_i)) \tag{5.116}$$

where $\widehat{\xi}_{(-i)}(x)$ is the estimator obtained by leaving out (x_i, Y_i). Unfortunately, there is no identity as in Theorem 5.34. There is, however, the following approximation from Loader (1999a). Recall the definition of $\ell(x, \xi)$ from (5.114) and let $\dot{\ell}(y, \xi)$ and $\ddot{\ell}(y, \xi)$ denote first and second derivatives of $\ell(y, \xi)$ with respect to ξ. Thus,

$$\dot{\ell}(y, \xi) = y - p(\xi)$$
$$\ddot{\ell}(y, \xi) = -p(\xi)(1 - p(\xi))$$

where $p(\xi) = e^{\xi}/(1 + e^{\xi})$. Define matrices X_x and W_x as in (5.54) and let V_x be a diagonal matrix with j^{th} diagonal entry equal to $-\ddot{\ell}(Y_i, \widehat{a}_0 + \widehat{a}_1(x_j - x_i))$. Then,

$$\text{CV} \approx \ell_x(\widehat{a}) + \sum_{i=1}^{n} m(x_i)\left(\dot{\ell}(Y_i, \widehat{a}_0)\right)^2 \tag{5.117}$$

where

$$m(x) = K(0)e_1^T (X_x^T W_x V_x X_x)^{-1} e_1 \tag{5.118}$$

and $e_1 = (1, 0, \ldots, 0)^T$. The effective degrees of freedom is

$$\nu = \sum_{i=1}^{n} m(x_i)\mathbb{E}(-\ddot{\ell}(Y_i, \widehat{a}_0)).$$

5.119 Example. Figure 5.12 shows the local linear logistic regression estimator for an example generated from the model $Y_i \sim \text{Bernoulli}(r(x_i))$ with $r(x) = e^{3\sin(x)}/(1 + e^{3\sin(x)})$. The solid line is the true function $r(x)$. The dashed line is the local linear logistic regression estimator. We also computed the local linear regression estimator which ignores the fact that the data are Bernoulli. The dotted line is the resulting local linear regression estimator.[9] cross-validation was used to select the bandwidth in both cases. ∎

5.120 Example. We introduced the BPD data in Example 4.6. The outcome Y is presence or absence of BPD and the covariate is $x = $ birth weight. The estimated logistic regression function (solid line) $r(x; \widehat{\beta}_0, \widehat{\beta}_1)$ together with the

---

[9]It might be appropriate to use a weighted fit since the variance of the Bernoulli is a function of the mean.

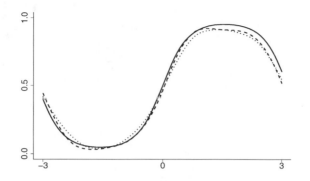

FIGURE 5.12. Local linear logistic regression. The solid line is the true regression function $r(x) = \mathbb{P}(Y = 1|X = x)$. The dashed line is the local linear logistic regression estimator. The dotted line is the local linear regression estimator.

data are shown in Figure 5.13. Also shown are two nonparametric estimates. The dashed line is the local likelihood estimator. The dotted line is the local linear estimator which ignores the binary nature of the Y_i's. Again we see that there is not a dramatic difference between the local logistic model and the local linear model. ∎

5.11   Scale-Space Smoothing

There is another approach to smoothing championed by Chaudhuri and Marron (1999) and Chaudhuri and Marron (2000) called **scale-space smoothing** that eschews the idea of selecting a single bandwidth. Let $\widehat{r}_h(x)$ denote an estimator using bandwidth h. The idea is to regard $\widehat{r}_h(x)$ as an estimator of $r_h(x) \equiv \mathbb{E}(\widehat{r}_h(x))$, as we did in Section 5.7. But rather than choosing a single bandwidth, we examine $\widehat{r}_h$ over a set of bandwidths h as a way of exploring the **scale-space surface**

$$\mathcal{S} = \left\{ r_h(x), x \in \mathcal{X}, h \in \mathcal{H} \right\}$$

where $\mathcal{X}$ is the range of x and $\mathcal{H}$ is the range of h.

One way to summarize the estimated scale-space surface

$$\widehat{\mathcal{S}} = \left\{ \widehat{r}_h(x), x \in \mathcal{X}, h \in \mathcal{H} \right\}$$

is to isolate important shape summaries. For example, Chaudhuri and Marron (1999) look for points x where $r_h'(x) = 0$ by using $\widehat{r}_h'(x)$ as a set of test

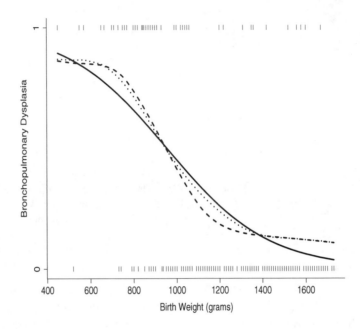

FIGURE 5.13. The BPD data. The data are shown with small vertical lines. The estimates are from logistic regression (solid line), local likelihood (dashed line) and local linear regression (dotted line).

statistics. They call the resulting method SiZer (significant zero crossings of derivatives).

5.12 Multiple Regression

Suppose now that the covariate is d-dimensional,

$$x_i = (x_{i1}, \ldots, x_{id})^T.$$

The regression equation takes the form

$$Y = r(x_1, \ldots, x_d) + \epsilon. \tag{5.121}$$

In principle, all the methods we have discussed carry over to this case easily. Unfortunately, the risk of a nonparametric regression estimator increases rapidly with the dimension d. This is the curse of dimensionality that we discussed in Section 4.5. It is worth revisiting this point now. In a one-dimensional problem, the optimal rate of convergence of a nonparametric estimator is $n^{-4/5}$ if r is assumed to have an integrable second derivative. In d dimensions the optimal rate of convergence is is $n^{-4/(4+d)}$. Thus, the sample size m required for a d-dimensional problem to have the same accuracy as a sample size n in a one-dimensional problem is $m \propto n^{cd}$ where $c = (4 + d)/(5d) > 0$. This implies the following fact:

> To maintain a given degree of accuracy of an estimator, the sample size must increase exponentially with the dimension d.

Put another way, confidence bands get very large as the dimension d increases. Nevertheless, let us press on and see how we might estimate the regression function.

LOCAL REGRESSION. Consider local linear regression. The kernel function K is now a function of d variables. Given a nonsingular positive definite $d \times d$ bandwidth matrix H, we define

$$K_H(x) = \frac{1}{|H|^{1/2}} K(H^{-1/2}x).$$

Often, one scales each covariate to have the same mean and variance and then we use the kernel

$$h^{-d} K(||x||/h)$$

where K is any one-dimensional kernel. Then there is a single bandwidth parameter h. At a target value $x = (x_1, \ldots, x_d)^T$, the local sum of squares is given by

$$\sum_{i=1}^{n} w_i(x) \left(Y_i - a_0 - \sum_{j=1}^{d} a_j(x_{ij} - x_j) \right)^2 \tag{5.122}$$

where

$$w_i(x) = K(||x_i - x||/h).$$

The estimator is

$$\widehat{r}_n(x) = \widehat{a}_0 \tag{5.123}$$

where $\widehat{a} = (\widehat{a}_0, \ldots, \widehat{a}_d)^T$ is the value of $a = (a_0, \ldots, a_d)^T$ that minimizes the weighted sums of squares. The solution $\widehat{a}$ is

$$\widehat{a} = (X_x^T W_x X_x)^{-1} X_x^T W_x Y \tag{5.124}$$

where

$$X_x = \begin{pmatrix} 1 & x_{11} - x_1 & \cdots & x_{1d} - x_d \\ 1 & x_{21} - x_1 & \cdots & x_{2d} - x_d \\ \vdots & \vdots & \ddots & \vdots \\ 1 & x_{n1} - x_1 & \cdots & x_{nd} - x_d \end{pmatrix}$$

and W_x is the diagonal matrix whose (i, i) element is $w_i(x)$.

The theoretical properties of local polynomial regression in higher dimensions is discussed in Ruppert and Wand (1994). The main result is as follows.

5.125 Theorem (Ruppert and Wand, 1994). *Let $\widehat{r}_n$ be the multivariate local linear estimator with bandwidth matrix H and assume the regularity conditions given in Ruppert and Wand (1994). Suppose that x is a nonboundary point. Conditional on $X_1, \ldots, X_n$ we have the following: The bias of $\widehat{r}_n(x)$ is*

$$\frac{1}{2}\mu_2(K)\text{trace}(H\mathcal{H}) + o_P(\text{trace}(H)) \tag{5.126}$$

where $\mathcal{H}$ is the matrix of second partial derivatives of r evaluated at x and $\mu_2(K)$ is the scalar defined by the equation $\int uu^T K(u)du = \mu_2(K)I$. The variance of $\widehat{r}_n(x)$ is

$$\frac{\sigma^2(x) \int K(u)^2 du}{n|H|^{1/2}f(x)}(1 + o_P(1)). \tag{5.127}$$

Also, the bias at the boundary is the same order as in the interior, namely, $O_P(\text{trace}(H))$.

Thus we see that in higher dimensions, local linear regression still avoids excessive boundary bias and design bias.

SPLINES. If we take a spline approach, we need to define splines in higher dimensions. For $d = 2$ we minimize

$$\sum_i (Y_i - \widehat{r}_n(x_{i1}, x_{i2}))^2 + \lambda J(r)$$

where

$$J(r) = \int\int \left[\left(\frac{\partial^2 r(x)}{\partial x_1^2}\right) + 2\left(\frac{\partial^2 r(x)}{\partial x_1 \partial x_2}\right) + \left(\frac{\partial^2 r(x)}{\partial x_2^2}\right) \right] dx_1 dx_2.$$

The minimizer $\widehat{r}_n$ is called a **thin-plate spline**. It is hard to describe and even harder (but certainly not impossible) to fit. See Green and Silverman (1994) for more details.

ADDITIVE MODELS. Interpreting and visualizing a high-dimensional fit is difficult. As the number of covariates increases, the computational burden becomes prohibitive. Sometimes, a more fruitful approach is to use an **additive model**. An additive model is a model of the form

$$Y = \alpha + \sum_{j=1}^{d} r_j(x_j) + e \tag{5.128}$$

where $r_1, \ldots, r_d$ are smooth functions. The model (5.128) is not identifiable since we can add any constant to α and subtract the same constant from one of the r_j's without changing the regression function. This problem can be fixed in a number of ways, perhaps the easiest being to set $\widehat{\alpha} = \overline{Y}$ and then regard the r_j's as deviations from $\overline{Y}$. In this case we require that $\sum_{i=1}^{n} \widehat{r}_j(x_i) = 0$ for each j.

The additive model is clearly not as general as fitting $r(x_1, \ldots, x_d)$ but it is much simpler to compute and to interpret and so it is often a good starting point. This is a simple algorithm for turning any one-dimensional regression smoother into a method for fitting additive models. It is called **backfitting**.

The Backfitting Algorithm

Initialization: set $\widehat{\alpha} = \overline{Y}$ and set initial guesses for $\widehat{r}_1, \ldots, \widehat{r}_d$.

Iterate until convergence: for $j = 1, \ldots, d$:

- Compute $\widetilde{Y}_i = Y_i - \widehat{\alpha} - \sum_{k \neq j} r_k(x_i)$, $i = 1, \ldots, n$.

- Apply a smoother to $\widetilde{Y}_i$ on x_j to obtain $\widehat{r}_j$.

- Set $\widehat{r}_j(x)$ equal to $\widehat{r}_j(x) - n^{-1} \sum_{i=1}^{n} \widehat{r}_j(x_i)$.

5.129 Example. Let us revisit Example 4.7 involving three covariates and one response variable. The data are plotted in Figure 5.14. Recall that the data are 48 rock samples from a petroleum reservoir, the response is permeability (in milli-Darcies) and the covariates are: the area of pores (in pixels out of 256 by 256), perimeter in pixels and shape (perimeter/$\sqrt{\text{area}}$). The goal is to predict permeability from the three covariates. First we fit the additive model

$$\text{permeability} = r_1(\text{area}) + r_2(\text{perimeter}) + r_3(\text{shape}) + \epsilon.$$

We could scale each covariate to have the same variance and then use a common bandwidth for each covariate. Instead, I took the more adventurous approach of performing cross-validation to choose a bandwidth h_j for covariate

x_j during each iteration of backfitting. I am not aware of any theory that guarantees convergence if the smoothing parameters are changed this way during the algorithm. Nonetheless, the bandwidths and the functions estimates converged rapidly. The estimates of r_1, r_2 and r_3 are shown in Figure 5.15. $\overline{Y}$ was added to each function before plotting it. Next consider a three-dimensional local linear fit (5.123). After scaling each covariate to have mean 0 and variance 1, we found that the bandwidth $h \approx 3.2$ minimized the cross-validation score. The residuals from the additive model and the full three-dimensional local linear fit are shown in Figure 5.16. Apparently, the fitted values are quite similar suggesting that the generalized additive model is adequate. ∎

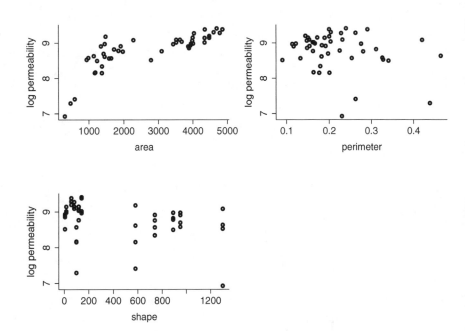

FIGURE 5.14. The rock data.

PROJECTION PURSUIT. Friedman and Stuetzle (1981) introduced another method for dealing with high-dimensional regression called **projection pursuit regression**. The idea is to approximate the regression function

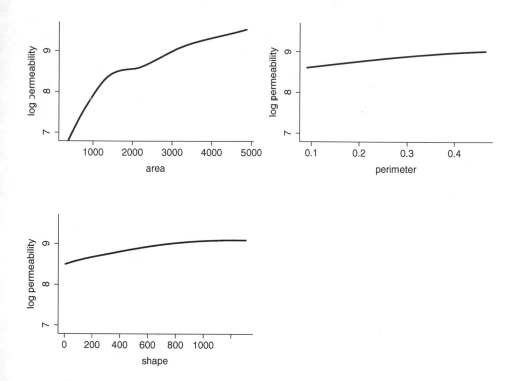

FIGURE 5.15. The rock data. The plots show $\widehat{r}_1$, $\widehat{r}_2$, and $\widehat{r}_3$ for the additive model $Y = \widehat{r}_1(x_1) + \widehat{r}_2(x_2) + \widehat{r}_3(x_3) + \epsilon$.

$r(x_1, \ldots, x_p)$ with a function of the form

$$\mu + \sum_{m=1}^{M} r_m(z_m)$$

where

$$z_m = \alpha_m^T x$$

and each α_m is a unit vector (length one) for $m = 1, \ldots, M$. Note that each z_m is the projection of x into a subspace. The direction vector α is chosen at each stage to minimize the fraction of unexplained variance. In more detail, let $S(\cdot)$ denote the mapping that outputs n fitted values from some smoothing method, given the Y_is and some one-dimensional covariate values $z_1, \ldots, z_n$. Let $\widehat{\mu} = \overline{Y}$ and replace Y_i with $Y_i - \overline{Y}$. Hence, the Y_i's now have mean 0. Similarly, scale the covariates so that they each have the same variance. Then do the following steps:

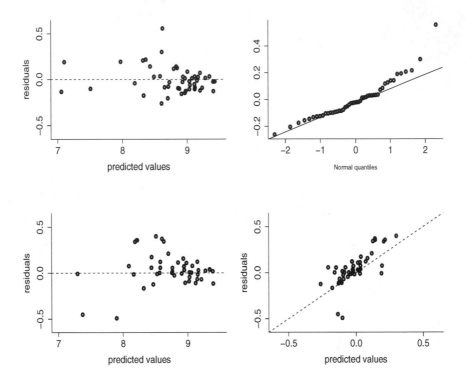

FIGURE 5.16. The residuals for the rock data. Top left: residuals from the additive model. Top right: qq-plot of the residuals from the additive model. Bottom left: residuals from the multivariate local linear model. Bottom right: residuals from the two fits plotted against each other.

Step 1: Initialize residuals $\widehat{\epsilon}_i = Y_i$, $i = 1, \ldots, n$ and set $m = 0$.
Step 2: Find the direction (unit vector) α that maximizes

$$I(\alpha) = 1 - \frac{\sum_{i=1}^n (\widehat{\epsilon}_i - S(\alpha^T x_i))^2}{\sum_{i=1}^n \widehat{\epsilon}_i^2}$$

and set $z_{mi} = \alpha^T x_i$, $\widehat{r}_m(z_{mi}) = S(z_{mi})$.
Step 3: Set $m = m + 1$ and update the residuals:

$$\widehat{\epsilon}_i \longleftarrow \widehat{\epsilon}_i - \widehat{r}_m(z_{mi}).$$

If $m = M$ stop, otherwise go back to Step 2.

5.130 Example. If we apply projection pursuit regression to the rock data with $M = 3$ we get the functions $\widehat{r}_1, \widehat{r}_2, \widehat{r}_3$ shown in 5.17. The fitting was done using the `ppr` command in R and each fit was obtained using smoothing splines where the smoothing parameter is chosen by generalized cross-validation. The direction vectors are

$$\alpha_1 = (.99, .07, .08)^T, \ \alpha_2 = (.43, .35, .83)^T, \ \alpha_3 = (.74, -.28, -.61)^T$$

Thus, $z_1 = .99\,\text{area} + .07\,\text{peri} + .08\,\text{shape}$ and so on. If we keep adding terms to the model, the residual sums of squares will keep getting smaller. The bottom left plot in Figure 5.17 shows the residual sums of squares as a function of the number of terms M. We see that after including one or two terms in the model, further terms add little. We could try to choose an optimal M by using cross-validation. ∎

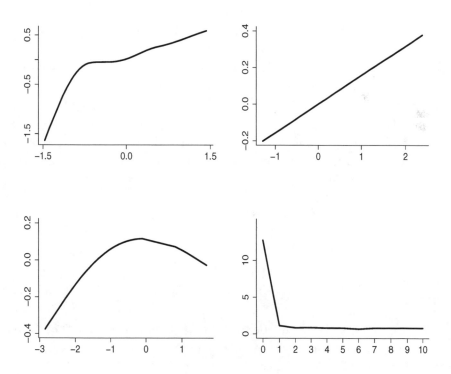

FIGURE 5.17. Projection pursuit applied to the rock data. The plots show $\widehat{r}_1$, $\widehat{r}_2$, and $\widehat{r}_3$.

REGRESSION TREES. A regression tree is a model of the form

$$r(x) = \sum_{m=1}^{M} c_m I(x \in R_m) \tag{5.131}$$

where $c_1, \ldots, c_M$ are constants and $R_1, \ldots, R_M$ are disjoint rectangles that partition the space of covariates. Tree models were introduced by Morgan and Sonquist (1963) and Breiman et al. (1984). The model is fitted in a recursive manner that can be represented as a tree; hence the name. Our description follows Section 9.2 of Hastie et al. (2001).

Denote a generic covariate value by $x = (x_1, \ldots, x_j, \ldots, x_d)$. The covariate for the ith observation is $x_i = (x_{i1}, \ldots, x_{ij}, \ldots, x_{id})$. Given a covariate j and a split point s we define the rectangles $R_1 = R_1(j, s) = \{x : x_j \leq s\}$ and $R_2 = R_2(j, s) = \{x : x_j > s\}$ where, in this expression, x_j refers the the jth covariate not the jth observation. Then we take c_1 to be the average of all the Y_i's such that $x_i \in R_1$ and c_2 to be the average of all the Y_i's such that $x_i \in R_2$. Notice that c_1 and c_2 minimize the sums of squares $\sum_{x_i \in R_1}(Y_i - c_1)^2$ and $\sum_{x_i \in R_2}(Y_i - c_2)^2$. The choice of which covariate x_j to split on and which split point s to use is based on minimizing the residual sums if squares. The splitting process is on repeated on each rectangle R_1 and R_2.

Figure 5.18 shows a simple example of a regression tree; also shown are the corresponding rectangles. The function estimate $\widehat{r}$ is constant over the rectangles.

Generally one grows a very large tree, then the tree is pruned to form a subtree by collapsing regions together. The size of the tree is a tuning parameter chosen as follows. Let N_m denote the number of points in a rectangle R_m of a subtree T and define

$$c_m = \frac{1}{N_m} \sum_{x_i \in R_m} Y_i, \quad Q_m(T) = \frac{1}{N_m} \sum_{x_i \in R_m} (Y_i - c_m)^2.$$

Define the complexity of T by

$$C_\alpha(T) = \sum_{m=1}^{|T|} N_m Q_m(T) + \alpha |T| \tag{5.132}$$

where $\alpha > 0$ and $|T|$ is the number of terminal nodes of the tree. Let T_α be the smallest subtree that minimizes C_α. The value $\widehat{\alpha}$ of α can be chosen by cross-validation. The final estimate is based on the tree $T_{\widehat{\alpha}}$.

5.133 Example. Figure 5.19 shows a tree for the rock data. Notice that the variable shape does not appear in the tree. This means that the shape variable

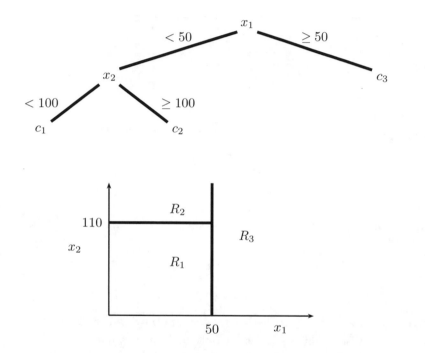

FIGURE 5.18. A regression tree for two covariates x_1 and x_2. The function estimate is $\hat{r}(x) = c_1 I(x \in R_1) + c_2 I(x \in R_2) + c_3 I(x \in R_3)$ where R_1, R_2 and R_3 are the rectangles shown in the lower plot.

was never the optimal covariate to split on in the algorithm. The result is that tree only depends on area and peri. This illustrates an important feature of tree regression: it automatically performs variable selection in the sense that a covariate x_j will not appear in the tree if the algorithm finds that the variable is not important. ∎

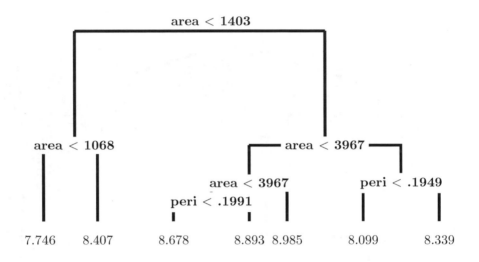

FIGURE 5.19. Regression tree for the rock data.

MARS. Regression trees are discontinuous and they do not easily fit main effects. (In contrast, additive models fit only main effects unless interactions are put in the model.) MARS—which stands for **multivariate adaptive regression splines**—was introduced by Friedman (1991) as an attempt to improve on regression trees.

The MARS algorithm is as follows. We follow Section 9.4 of Hastie et al. (2001). Define

$$\ell(x, j, t) = (x_j - t)I(x_j > t), \quad r(x, j, t) = (t - x_j)I(x_j < t).$$

Both $\ell(x, j, t)$ and $r(x, j, t)$ are functions of the whole vector $x = (x_1, \ldots, x_d)$ but their values only depend on the jth component x_j. Let

$$\mathcal{C}_j = \Big\{ \ell(x, j, t), r(x, j, t), t \in \{x_{1j}, \ldots, x_{nj}\} \Big\}.$$

Thus, $\mathcal{C}_j$ is a set of linear splines that depend only on x_j, with knots at the observations. Now let $\mathcal{C} = \bigcup_{j=1}^d \mathcal{C}_j$. A MARS model is of the form

$$r(x) = \beta_0 + \sum_{m=1}^M \beta_m h_m(x) \tag{5.134}$$

where every function h_m is either in $\mathcal{C}$ or is the product of two or more such functions. The model is fitted in a forward, stepwise manner, much like regression trees. See Hastie et al. (2001) for more details.

TENSOR PRODUCT MODELS. Another class of models for multiple regression are of the form

$$r(x) = \sum_{m=1}^M \beta_m h_m(x) \tag{5.135}$$

where each h_m is a basis function in a tensor product space. These models will be considered in Chapter 8.

5.13   Other Issues

Here we discuss a few other issues related to nonparametric regression.

PLUG-IN BANDWIDTHS. An alternative method to cross-validation for choosing the bandwidth is to use a **plug-in bandwidth**. The idea is to write down a formula for the asymptotically optimal bandwidth and then insert estimates of the unknown quantities in the formula. We describe one possible approach based on Section 4.2 of Fan and Gijbels (1996).

When using local linear regression, and assuming that the X_i's are randomly selected from some density $f(x)$, the (asymptotically) optimal bandwidth is

$$h_* = \left(\frac{C \int \frac{\sigma^2(x)}{f(x)} dx}{n \int (r^{(2)}(x))^2 dx} \right)^{1/5} \tag{5.136}$$

where

$$C = \frac{\int K^2(t) dt}{(\int K(t) dt)^2} \tag{5.137}$$

and $r^{(2)}$ is the second derivative of r. To get a crude estimate of h_* proceed as follows. Fit a global quartic

$$\tilde{r}(x) = \hat{\beta}_0 + \hat{\beta}_1 x + \hat{\beta}_2 x^2 + \hat{\beta}_3 x^3 + \hat{\beta}_4 x^4$$

using least squares and let $\tilde{\sigma}^2 = n^{-1} \sum_{i=1}^{n} (Y_i - \tilde{r}(x_i))^2$. We call $\tilde{r}(x)$ the **pilot estimate**. Let (a, b) denote the range of the X_i's and approximate f with a uniform over (a, b). Then,

$$
\begin{aligned}
n \int_a^b r^{(2)}(x)^2 dx &= n \int_a^b \frac{r^{(2)}(x)^2}{f(x)} f(x) dx \\
&\approx \sum_{i=1}^{n} \frac{r^{(2)}(X_i)^2}{f(X_i)} = (b-a) \sum_{i=1}^{n} (r^{(2)}(X_i))^2
\end{aligned}
$$

and we estimate h_* by

$$
h_* \approx \left(\frac{C \tilde{\sigma}^2 (b-a)}{\sum_{i=1}^{n} (\tilde{r}^{(2)}(X_i))^2} \right)^{1/5}. \tag{5.138}
$$

TESTING FOR LINEARITY. A nonparametric estimator $\hat{r}_n$ can be used to construct a test to see whether a linear fit is adequate. Consider testing

$$
H_0 : r(x) = \beta_0 + \beta_1 x \text{ for some } \beta_0, \beta_1
$$

versus the alternative that H_0 is false.

Denote the hat matrix from fitting the linear model by H and the smoothing matrix from fitting the nonparametric regression by L. Let

$$
T = \frac{\|LY - HY\|/\lambda}{\hat{\sigma}^2}
$$

where $\lambda = \text{tr}((L - H)^T (L - H))$ and $\hat{\sigma}^2$ is defined by (5.86). Loader (1999a) points out that under H_0, the F-distribution with ν and $n - 2\nu_1 + \nu_2$ degrees of freedom provides a rough approximation to the distribution of T. Thus we would reject H_0 at level α if $T > F_{\nu, n-2\nu_1+\nu_2, \alpha}$. A more rigorous test that uses bootstrapping to estimate the null distribution is described in Härdle and Mammen (1993).

As with any test, the failure to reject H_0 should not be regarded as proof that H_0 is true. Rather, it indicates that the data are not powerful enough to detect deviations from H_0. In such cases, a linear fit might be considered a reasonable tentative model. Of course, making such decisions based solely on the basis of a test can be dangerous.

OPTIMALITY. Local linear estimators have certain optimality properties. We highlight a few results from Fan and Gijbels (1996). Let x_0 be an interior (nonboundary) point and let

$$\mathcal{F} = \{r : \ |r(x) - r(x_0) - (x - x_0)r'(x_0)| \leq C|x - x_0|\}.$$

Suppose that the covariates X is random with density f that is positive at x_0. Also, assume that the variance function $\sigma(x)$ is continuous at x_0. Let $\mathcal{L}$ denote all linear estimators of $r(x_0)$. The **linear minimax risk** is defined by

$$R_n^L = \inf_{\widehat{\theta} \in \mathcal{L}} \sup_{r \in \mathcal{F}} \mathbb{E}((\widehat{\theta} - r(x_0))^2 | X_1, \ldots, X_n). \tag{5.139}$$

Fan and Gijbels (1996) show that

$$R_n^L = \frac{3}{4} 15^{-1/5} \left(\frac{\sqrt{C}\sigma^2(x_0)}{nf(x_0)} \right)^{4/5} (1 + o_P(1)). \tag{5.140}$$

Moreover, this risk is achieved by the local linear estimator $\widehat{r}_*$ with Epanechnikov kernel and bandwidth

$$h_* = \left(\frac{15\sigma^2(x_0)}{f(x_0)C^2 n} \right)^{1/5}.$$

The **minimax risk** is defined by

$$R_n = \inf_{\widehat{\theta}} \sup_{r \in \mathcal{F}} \mathbb{E}((\widehat{\theta} - r(x_0))^2 | X_1, \ldots, X_n) \tag{5.141}$$

where the infimum is now over all estimators. Fan and Gijbels (1996) show that $\widehat{r}_*$ is nearly minimax in the sense that

$$\frac{R_n}{\sup_{r \in \mathcal{F}} \mathbb{E}((\widehat{r}^*(x_0) - r(x_0))^2 | X_1, \ldots, X_n)} \geq (0.894)^2 + o_P(1). \tag{5.142}$$

See Chapter 7 for more discussion on minimaxity.

DERIVATIVE ESTIMATION. Suppose we want to estimate the kth derivative $r^{(k)}(x)$ of $r(x)$. Recall that the local polynomial estimator begins with the approximation

$$r(u) \approx a_0 + a_1(u - x) + \frac{a_2}{2}(u - x)^2 + \cdots + \frac{a_p}{p!}(u - x)^p.$$

Thus, $r^{(k)}(x) \approx a_k$ and we can estimate this with

$$\widehat{r}_n^{(k)}(x) = \widehat{a}_k = \sum_{i=1}^{n} \ell_i(x, k) Y_i \tag{5.143}$$

where $\ell(x, k)^T = (\ell_1(x, k), \ldots, \ell_n(x, k))$,

$$\ell(x, k)^T = e_{k+1}^T (X_x^T W_x X_x)^{-1} X_x^T W_x,$$

$$e_{k+1} = (\underbrace{0, \ldots, 0}_{k}, 1, \underbrace{0, \ldots, 0}_{p-k})^T$$

and X_x and W_x are defined in (5.54).

Warning! Note that $\widehat{r}_n^{(k)}(x)$ is not equal to the kth derivative of $\widehat{r}_n$.

To avoid boundary and design bias, take the order of the polynomial p so that $p - k$ is odd. A reasonable default is $p = k + 1$. So, for estimating the first derivative, we would use local quadratic regression instead of local linear regression. The next theorem gives the large sample behavior of $\widehat{r}_n^{(k)}$. A proof can be found in Fan (1992) and Fan and Gijbels (1996). To state the theorem, we need a few definitions. Let $\mu_j = \int u^j K(u) du$ and $\nu_j = \int u^j K^2(u) du$. Define $(p + 1) \times (p + 1)$ matrices S and S^* whose (r, s) entries are

$$S_{rs} = \mu_{r+s-2}, \quad S_{rs}^* = \nu_{r+s-2}.$$

Also, let $c_p = (\mu_{p+1}, \ldots, \mu_{2p+1})^T$ and $\widetilde{c}_p = (\mu_{p+2}, \ldots, \mu_{2p+2})^T$. Finally, let

$$e_{k+1} = (\underbrace{0, \ldots, 0}_{k}, 1, \underbrace{0, \ldots, 0}_{p-k})^T.$$

5.144 Theorem. *Let $Y_i = r(X_i) + \sigma(X_i)\epsilon_i$ for $i = 1, \ldots, n$. Assume that $X_1, \ldots, X_n$ are a sample from a distribution with density f and that (i) $f(x) > 0$, (ii) f, $r^{(p+1)}$ and σ^2 are continuous in a neighborhood of x, and (iii) $h \to 0$ and $nh \to \infty$. Then, given $X_1, \ldots, X_n$, we have the following:*

$$\mathbb{V}(\widehat{r}_n^{(k)}(x)) = e_{k+1}^T S^{-1} S^* S^{-1} e_{k+1} \frac{k!^2 \sigma^2(x)}{f(x)nh^{1+2k}} + o_P\left(\frac{1}{nh^{1+2k}}\right). \quad (5.145)$$

If $p - k$ is odd, the bias is

$$\mathbb{E}(\widehat{r}_n^{(k)}(x)) - r(x) = e_{k+1}^T S^{-1} c_p \frac{k!}{(p+1)!} r^{(p+1)}(x) h^{p+1-k}$$

$$+ o_P(h^{p+1-k}). \quad (5.146)$$

If $p - k$ is even then f' and $m^{(p+2)}$ are continuous in a neighborhood of x and $nh^3 \to \infty$ then the bias is

$$\mathbb{E}(\widehat{r}_n^{(k)}(x)) - r(x) = e_{k+1}^T S^{-1} \widetilde{c}_p \frac{k!}{(p+2)!}$$

$$\times \left(r^{(p+2)}(x) + (p+2)m^{(p+1)}(x)\frac{f'(x)}{f(x)} \right) h^{p+2-k}$$

$$+ o_P(h^{p+2-k}). \quad (5.147)$$

Define

$$K_k^*(t) = K(t) \sum_{\ell=0}^{p} S^{(k-1)(\ell-1)} t^\ell.$$

Then it can be shown that the (asymptotically) optimal bandwidth is

$$h_* = \left(\frac{C(k,p) \int \frac{\sigma^2(x)}{f(x)} dx}{n \int (r^{(p+1)}(x))^2 dx} \right)^{1/(2p+3)} \tag{5.148}$$

where

$$C(k,p) = \left(\frac{(p+1)!^2 (2k+1) \int K_k^{*2}(t) dt}{2(p+1-k)(\int t^{p+1} K_k^*(t) dt)^2} \right)^{1/(2p+3)}.$$

Estimating a derivative is much harder then estimating the regression function because we observe the regression function (plus error) but we do not observe the derivative directly. See Chapter 6.1 of Loader (1999a) for a cogent discussion of the difficulties in estimating derivatives.

VARIABLE BANDWIDTHS AND ADAPTIVE ESTIMATION. Instead of using one bandwidth h, we might try to use a bandwidth $h(x)$ that varies with x. Choosing a bandwidth this way is called **variable bandwidth selection**. This seems appealing because it might allow us to **adapt** to varying degrees of smoothness. For example, $r(x)$ might be spatially inhomogeneous meaning that it is smooth for some values of x and wiggly for other values of x. Perhaps we should use a large bandwidth for the smooth regions and a small bandwidth for the wiggly regions. Such a procedure is said to be **locally adaptive** or **spatially adaptive**. See Chapter 4 of Fan and Gijbels (1995) and Ruppert (1997), for example. However, the improvements in the function estimate are often quite modest unless the sample size is very large and the noise level is low. For more discussion on spatial adaptation, see Chapter 9 and in particular Section 9.9.

CORRELATED DATA. We have assumed that the errors $\epsilon_i = Y_i - r(x_i)$ are independent. When there is dependence between the errors, the methods need to be modified. The type of modification required depends on the type of dependence that is present. For time-ordered data, for example, time series methods are usually required. More generally, some knowledge of the dependence structure is required to devise suitable estimation methods. See Chapter 10 for more on this point.

ROBUSTNESS AND QUANTILE REGRESSION. The estimators we have used are based on squared error loss. This is an easy loss function to use but the resulting estimator is potentially not robust to outliers. In **robust regression**, we choose $\widehat{a}$ to minimize

$$\sum_{i=1}^{n} w_i(x)\rho\left(\frac{\left(Y_i - a_0 - a_1(u-x) + \cdots + \frac{a_p}{p!}(u-x)^p\right)}{s}\right) \qquad (5.149)$$

instead of minimizing (5.52). Here, s is some estimate of the standard deviation of the residuals. Taking $\rho(t) = t^2$ gets us back to squared error loss. A more robust estimator is obtained by using **Huber's function** defined by the equation

$$\rho'(t) = \max\{-c, \min(c, t)\} \qquad (5.150)$$

where c is a tuning constant. We get back squared error as $c \to \infty$ and we get absolute error as $c \to 0$. A common choice is $c = 4.685$ which provides a compromise between the two extremes. Taking

$$\rho(t) = |t| + (2\alpha - 1)t \qquad (5.151)$$

yields **quantile regression**. In this case, $\widehat{r}_n(x)$ estimates $\xi(x)$, where $\mathbb{P}(Y \leq \xi(x)|X = x) = \alpha$ so that $\xi(x)$ is the α quantile of the conditional distribution of Y given x. See Section 5.5 of Fan and Gijbels (1996) for more detail.

MEASUREMENT ERROR. In some cases we do not observe x directly. Instead, we observe a corrupted version of x. The observed data are $(Y_1, W_1), \ldots, (Y_n, W_n)$ where

$$\begin{aligned} Y_i &= r(x_i) + \epsilon_i \\ W_i &= x_i + \delta_i \end{aligned}$$

for some errors δ_i. This is called a **measurement error** problem or an **errors-in-variables** problem. Simply regressing the Y_i's on the W_i's leads to inconsistent estimates of $r(x)$. We will discuss measurement error in more detail in Chapter 10.

DIMENSION REDUCTION AND VARIABLE SELECTION. One way to deal with the curse of dimensionality is to try to find a low-dimension approximation

to the data. Techniques include **principal component analysis, independent component analysis projection pursuit** and others. See Hastie et al.
(2001) for an introduction to these methods as well as relevant references.

An alternative is to perform **variable selection** in which covariates that
do not predict Y well are removed from the regression. See Zhang (1991), for
example. Currently, there seems to be few variable selection methods in nonparametric regression that are both practical and have a rigorous theoretical
justification.

CONFIDENCE SETS FOR MULTIPLE REGRESSION. The confidence band
method in Section 5.7 can also be used for additive models. The method
also extends to linear smoothers in higher dimensions as explained in Sun and
Loader (1994). For more complicated methods like trees, MARS and projection pursuit regression, I am not aware of rigorous results that lead to valid
bands.

UNDERSMOOTHING. One approach to dealing with the bias problem in constructing confidence sets is undersmoothing. The issue is discussed in Hall
(1992b), Neumann (1995), Chen and Qin (2000), and Chen and Qin (2002).
Here, we briefly discuss the results of Chen and Qin.

Let $\widehat{r}_n(x)$ be the local linear estimator using bandwidth h and assume that
the kernel K has support on $[-1, 1]$. Let

$$\alpha_j(x/h) = \int_{-1}^{x/h} u^j K(u) du,$$

$$\widehat{f}_0(x) = \frac{1}{n} \sum_{i=1}^{n} \frac{1}{h} K\left(\frac{x - X_i}{h}\right), \quad \widehat{f}(x) = \widehat{f}_0(x)/\alpha_0(x/h),$$

$$\widehat{\sigma}^2(x) = \frac{\frac{1}{n} \sum_{i=1}^{n} \frac{1}{h} K\left(\frac{x-X_i}{h}\right) (Y_i - \widehat{r}_n(x))^2}{\widehat{f}_0(x)}$$

and

$$I(x) = \widehat{r}_n(x) \pm z_\alpha \sqrt{\frac{v(x/h)\widehat{\sigma}(x)}{nh\widehat{f}(x)}}$$

where

$$v(x/h) = \frac{\int_{-1}^{x/h} (\alpha_2(x/h) - u\alpha_1(x/h))^2 K^2(u) du}{(\alpha_0(x/h)\alpha_2(x/h) - \alpha_1^2(x/h))^2}.$$

Undersmoothing eliminates the asymptotic bias; this means we must take $nh^5 \to 0$. Assuming we do take $nh^5 \to 0$, and subject to some regularity conditions, Chen and Qin (2002) show that

$$\mathbb{P}(r(x) \in I(x)) = 1 - \alpha + O\left(nh^5 + h^2 + \frac{1}{nh}\right) \qquad (5.152)$$

at interior points and

$$\mathbb{P}(r(x) \in I(x)) = 1 - \alpha + O\left(nh^5 + h + \frac{1}{nh}\right) \qquad (5.153)$$

near the boundary. It is interesting that local linear regression eliminates boundary bias of $\hat{r}_n$ but the coverage probability is poor near the boundaries. The lack of uniformity of the accuracy of the coverage probability can be fixed using an approach in Chen and Qin (2000). The confidence interval they propose is

$$\{\theta : \ell(\theta) \leq c_\alpha\} \qquad (5.154)$$

where c_α is the upper α quantile of a χ_1^2 random variable,

$$\ell(\theta) = 2 \sum_{i=1}^{n} \log(1 + \lambda(\theta)W_i(Y_i - \theta)),$$

$\lambda(\theta)$ is defined by

$$\sum_{i=1}^{n} \frac{W_i(Y_i - \theta)}{1 + \lambda(\theta)W_i(Y_i - \theta)} = 0,$$

$$W_i = K\left(\frac{x - X_i}{h}\right)\left(s_{n,2} - \frac{(x - X_i)s_{n,1}}{h}\right),$$

and

$$s_{n,j} = \frac{1}{nh} \sum_{i=1}^{n} \frac{K\left(\frac{x - X_i}{h}\right)(x - X_i)^j}{h^j}.$$

Assuming we do take $hn^5 \to 0$, and subject to some regularity conditions, Chen and Qin (2000) show that

$$\mathbb{P}(r(x) \in I(x)) = 1 - \alpha + O\left(nh^5 + h^2 + \frac{1}{nh}\right) \qquad (5.155)$$

over all x. The optimal bandwidth, in terms of minimizing the coverage error is

$$h_* = \frac{c}{n^{1/3}}.$$

Unfortunately, the constant c depends on the unknown function r. It would appear that practical implementation is still an open research problem.

5.14   Bibliographic Remarks

The literature on nonparametric regression is very large. Some good starting points are Fan and Gijbels (1996), Härdle (1990), Loader (1999a), Hastie and Tibshirani (1999), and Hastie et al. (2001). A good source of the theory of splines is Wahba (1990). See also Hastie et al. (2001) and Ruppert et al. (2003). Local regression and local likelihood are discussed in detail in Loader (1999a) and Fan and Gijbels (1996). Variable bandwidth selection is discussed in Fan and Gijbels (1995).

5.15   Appendix

DERIVATION OF THE TUBE FORMULA (5.100). Let $W(x) = \sum_{i=1}^{n} Z_i T_i(x)$ and recall that $||T(x)||^2 = \sum_{i=1}^{n} T_i(x)^2 = 1$ so that the vector $T(x)$ is on the unit sphere, for each x. Since, $Z = (Z_1, \ldots, Z_n)$ are multivariate Normal,

$$
\begin{aligned}
\mathbb{P}(\sup_x W(x) > c) &= \mathbb{P}(\sup_x \langle Z, T(x) \rangle > c) \\
&= \mathbb{P}\left(\sup_x \left\langle \frac{Z}{||Z||}, T(x) \right\rangle > \frac{c}{||Z||} \right) \\
&= \int_{c^2}^{\infty} \mathbb{P}\left(\sup_x \langle U, T(x) \rangle > \frac{c}{\sqrt{y}} \right) h(y) dy
\end{aligned}
$$

where $U = (U_1, \ldots, U_n)$ is uniformly distributed on the $n-1$-dimensional unit sphere S and $h(y)$ is the density for a χ^2 with n degrees of freedom. Since $||U - T(x)||^2 = 2(1 - \langle U, T(x) \rangle)$, we see that $\sup_x \langle U, T(x) \rangle > \frac{c}{\sqrt{y}}$ if and only if $U \in \text{tube}(r, M)$ where $r = \sqrt{2(1 - c/\sqrt{y})}$, $M = \{T(x) : x \in \mathcal{X}\}$ is a manifold on the sphere S,

$$
\text{tube}(r, M) = \{u : d(u, M) \leq r\}
$$

and

$$
d(u, M) = \inf_{x \in \mathcal{X}} ||u - T(x)||.
$$

Therefore,

$$
\begin{aligned}
\mathbb{P}\left(\sup_x \langle U, T(x) \rangle > \frac{c}{\sqrt{y}} \right) &= \mathbb{P}(U \in \text{tube}(r, M)) \\
&= \frac{\text{volume}(\text{tube}(r, M))}{A_n}
\end{aligned}
$$

where $A_n = 2\pi^{n/2}/\Gamma(n/2)$ is the area of the unit sphere. The formula for volume$(\text{tube}(r, M))$ was derived by Hotelling (1939) and Naiman (1990) and is given by

$$\kappa_0 \frac{A_n}{A_2} \mathbb{P}(B_{1,(n-2)/2} \geq w^2) + \ell_0 \frac{A_n}{2A_1} \mathbb{P}(B_{1/2,(n-1)/2} \geq w^2)$$

where $w = c/\sqrt{y}$. Inserting this into the integral and ignoring terms of order smaller than $c^{-1/2} e^{-c^2/2}$ yields (5.100).

The formula can also be obtained using the upcrossing theory of Rice (1939). Specifically, if W is a Gaussian process on $[0, 1]$ and if N_c denotes the number of upcrossings of W above c then

$$
\begin{aligned}
\mathbb{P}(\sup_x W(x) > c) &= \mathbb{P}(N_c \geq 1 \text{ or } W(0) > c) \\
&\leq \mathbb{P}(N_c \geq 1) + \mathbb{P}(W(0) > c) \\
&\leq \mathbb{E}(N_c) + \mathbb{P}(W(0) > c).
\end{aligned}
$$

Since $W(0)$ has a Normal distribution, the second term can be easily computed. Moreover, under smoothness conditions on W we have

$$\mathbb{E}(N_c) = \int_0^1 \int_0^\infty y p_t(c, y) dy dt \qquad (5.156)$$

where p_t is the density of $(W(t), W'(t))$.

5.16   Exercises

1. In Example 5.24, construct the smoothing matrix L and verify that $\nu = m$.

2. Prove Theorem 5.34.

3. Get the data on fragments of glass collected in forensic work from the book website. Let Y be refractive index and let x be aluminium content (the fourth variable). Perform a nonparametric regression to fit the model $Y = r(x) + \epsilon$. Use the following estimators: (i) regressogram, (ii) kernel, (iii) local linear, (iv) spline. In each case, use cross-validation to choose the amount of smoothing. Estimate the variance. Construct 95 percent confidence bands for your estimates. Pick a few values of x and, for each value, plot the effective kernel for each smoothing method. Visually compare the effective kernels.

4. Get the motorcycle data from the book website. The covariate is time (in milliseconds) and the response is acceleration at time of impact. Use cross-validation to fit a smooth curve using local linear regression.

5. Show that with suitable smoothness assumptions on $r(x)$, $\widehat{\sigma}^2$ in equation (5.89) is a consistent estimator of σ^2.

6. Prove Theorem 5.34.

7. Prove Theorem 5.60.

8. Find conditions under which the estimate in equation (5.86) is consistent.

9. Consider the data in Exercise 3. Examine the fit as a function of the bandwidth h. Do this by plotting the fit for many values of h. Add confidence bands to all the fits. If you are feeling very ambitious, read Chaudhuri and Marron (1999) and apply that method.

10. Using five equally spaced knots on (0,1), construct a B-spline basis of order M for $M = 1, \ldots, 5$. Plot the basis functions.

11. Get the motorcycle data from the book website. Fit a cubic regression spline with equally spaced knots. Use leave-one-out cross-validation to choose the number of knots. Now fit a smoothing spline and compare the fits.

12. Recall the Doppler function defined in Example 5.63. Generate 1000 observations from the model $Y_i = r(x_i) + \sigma \epsilon_i$ where $x_i = i/n$ and $\epsilon_i \sim N(0, 1)$. Make three data sets corresponding to $\sigma = .1$, $\sigma = 1$ and $\sigma = 3$. Plot the data. Estimate the function using local linear regression. Plot the cross-validation score versus the bandwidth. Plot the fitted function. Find and plot a 95 percent confidence band.

13. Repeat the previous question but use smoothing splines.

14. Download the air quality data set from the book website. Model ozone as a function of temperature. Use kernel regression and compare the fit you get when you choose the bandwidth using cross-validation, generalized cross-validation, C_p and the plug-in method.

15. Let $Y_i \sim N(\mu_i, 1)$ for $i = 1, \ldots, n$ be independent observations. Find the estimators that minimizes each of the following penalized sums of squares:

$$(a) \qquad \sum_{i=1}^{n} (Y_i - \widehat{\mu}_i)^2 + \lambda \sum_{i=1}^{n} \widehat{\mu}_i^2$$

$$(b) \qquad \sum_{i=1}^{n}(Y_i - \widehat{\mu}_i)^2 + \lambda \sum_{i=1}^{n} |\widehat{\mu}_i|$$

$$(c) \qquad \sum_{i=1}^{n}(Y_i - \widehat{\mu}_i)^2 + \lambda \sum_{i=1}^{n} I(\widehat{\mu}_i = 0).$$

16. Show that a locally polynomial smoother of order p reproduces polynomials of order p.

17. Suppose that $r : [0, 1] \to \mathbb{R}$ satisfies the following **Lipschitz** condition:

$$\sup_{0 \leq x < y \leq 1} |r(y) - r(x)| \leq L|y - x| \qquad (5.157)$$

where $L > 0$ is given. The class of all such functions is denoted by $\mathcal{F}_{\text{lip}}(L)$. What is the maximum bias of a kernel estimator $\widehat{r}_n$ based on bandwidth h, if $r \in \mathcal{F}_{\text{lip}}(L)$?

18. Implement quantile regression on the glass data (Exercise 3) with $\alpha = 1/2$.

19. Prove that the weights $\ell_i(x)$ for the local polynomial smoother satisfy

$$\ell_i(x) = K\left(\frac{x_i - x}{h}\right) P_i(x) \qquad (5.158)$$

for some polynomial

$$P_i(x) = \alpha_0 + \alpha_1(x_i - x) + \cdots + \alpha_p(x_i - x)^p.$$

Moreover, if the ith observation (x_i, Y_i) is omitted, the resulting weights satisfy (5.32). Thus, while we took (5.32) as the definition of the leave-one-out weights, one can derive this form of the weights.

20. Suppose that $\ell_i(x) = K((x - x_i)/h)$ for some smooth kernel K and that the x_i's are equally spaced. Define κ_0 as in (5.101). Show that, if we ignore boundary effects,

$$\kappa_0 \approx \left(\frac{b - a}{h}\right) \frac{||K'||}{||K||}$$

where $||g||^2 = \int_a^b g^2(x)dx$.

21. Show how to construct a confidence band for the derivative estimator $\widehat{r}^{(k)}$ given in (5.143). *Hint:* Note that the estimator is linear and follow the construction of the confidence band for $\widehat{r}_n(x)$.

22. Download the air quality data set from the book website. Model ozone as a function of solar R, wind and temperature. Use (i) multiple local linear regression, (ii) projection pursuit, (iii) additive regression, (iv) regression trees, and (v) MARS. Compare the results.

23. Explain how to construct confidence bands in additive models. Apply this to the data from Exercise 22.

24. Let $\widehat{r}_n(x_1, x_2) = \sum_{i=1}^{n} Y_i \ell_i(x_1, x_2)$ be a linear estimator of the multiple regression function $r(x_1, x_2)$. Suppose we want to test the hypothesis that the covariate x_2 can be dropped from the regression. One possibility is to form a linear estimator of the form $\widetilde{r}_n(x_1) = \sum_{i=1}^{n} Y_i \widetilde{\ell}_i(x_1)$ and then compute

$$T = \sum_{i=1}^{n} (\widehat{r}_n(x_{1i}, x_{2i}) - \widetilde{r}_n(x_{1i}))^2.$$

(i) Assume that the true model is $Y_i = r(x_{1i}) + \epsilon_i$ where $\epsilon_i \sim N(0, \sigma^2)$. For simplicity take σ known. Find an expression for the distribution of T.

(ii) The null distribution in part (i) depends on the unknown function $r(x_1)$. How might you estimate the null distribution?

(iii) Create simulated data from the model in (i) (use any function $r(x_1)$ you like) and see if your proposed method in (ii) does approximate the null distribution.

6
Density Estimation

Let F be a distribution with probability density $f = F'$ and let

$$X_1, \ldots, X_n \sim F$$

be an IID sample from F. The goal of **nonparametric density estimation** is to estimate f with as few assumptions about f as possible. We denote the estimator by $\widehat{f}_n$. As with nonparametric regression, the estimator will depend on a smoothing parameter h and choosing h carefully is important.

6.1 Example (Bart Simpson). The top left plot in Figure 6.1 shows the density

$$f(x) = \frac{1}{2}\phi(x; 0, 1) + \frac{1}{10}\sum_{j=0}^{4} \phi(x; (j/2) - 1, 1/10) \tag{6.2}$$

where $\phi(x; \mu, \sigma)$ denotes a Normal density with mean μ and standard deviation σ. Marron and Wand (1992) call this density "the claw" although we will call it the Bart Simpson density. Based on 1000 draws from f, I computed a kernel density estimator, described later in the chapter. The top right plot is based on a small bandwidth h which leads to undersmoothing. The bottom right plot is based on a large bandwidth h which leads to oversmoothing. The bottom left plot is based on a bandwidth h which was chosen to minimize estimated risk. This leads to a much more reasonable density estimate. ∎

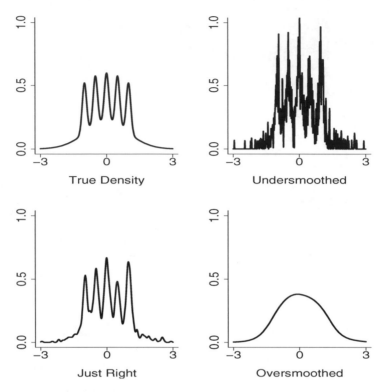

FIGURE 6.1. The Bart Simpson density from Example 6.1. Top left: true density. The other plots are kernel estimators based on $n = 1000$ draws. Bottom left: bandwidth $h = 0.05$ chosen by leave-one-out cross-validation. Top right: bandwidth $h/10$. Bottom right: bandwidth $10h$.

6.1   Cross-Validation

We will evaluate the quality of an estimator $\widehat{f}_n$ with the risk, or integrated mean squared error, $R = \mathbb{E}(L)$ where

$$L = \int (\widehat{f}_n(x) - f(x))^2 dx$$

is the integrated squared error loss function. The estimators will depend on some smoothing parameter h and we will choose h to minimize an estimate of the risk. The usual method for estimating risk is **leave-one-out cross-validation**. The details are different for density estimation than for regression. In the regression case, the cross-validation score was defined as $\sum_{i=1}^{n}(Y_i - \widehat{r}_{(-i)}(x_i))^2$ but in density estimation, there is no response variable Y. Instead, we proceed as follows.

The loss function, which we now write as a function of h, (since $\widehat{f}_n$ will depend on some smoothing parameter h) is

$$
\begin{aligned}
L(h) &= \int (\widehat{f}_n(x) - f(x))^2 \, dx \\
&- \int \widehat{f}_n^2(x) \, dx \quad 2 \int \widehat{f}_n(x) f(x) dx + \int f^2(x) \, dx.
\end{aligned}
$$

The last term does not depend on h so minimizing the loss is equivalent to minimizing the expected value of

$$
J(h) = \int \widehat{f}_n^2(x) \, dx - 2 \int \widehat{f}_n(x) f(x) dx. \tag{6.3}
$$

We shall refer to $\mathbb{E}(J(h))$ as the risk, although it differs from the true risk by the constant term $\int f^2(x) \, dx$.

6.4 Definition. *The* **cross-validation estimator of risk** *is*

$$
\widehat{J}(h) = \int \left(\widehat{f}_n(x) \right)^2 dx - \frac{2}{n} \sum_{i=1}^{n} \widehat{f}_{(-i)}(X_i) \tag{6.5}
$$

where $\widehat{f}_{(-i)}$ is the density estimator obtained after removing the ith observation. We refer to $\widehat{J}(h)$ as the cross-validation score or estimated risk.

6.2   Histograms

Perhaps the simplest nonparametric density estimator is the histogram. Suppose f has its support on some interval which, without loss of generality, we take to be $[0, 1]$. Let m be an integer and define **bins**

$$
B_1 = \left[0, \frac{1}{m} \right), \quad B_2 = \left[\frac{1}{m}, \frac{2}{m} \right), \quad \ldots, \quad B_m = \left[\frac{m-1}{m}, 1 \right]. \tag{6.6}
$$

Define the **binwidth** $h = 1/m$, let Y_j be the number of observations in B_j, let $\widehat{p}_j = Y_j/n$ and let $p_j = \int_{B_j} f(u) du$.

The **histogram estimator** is defined by

$$
\widehat{f}_n(x) = \sum_{j=1}^{m} \frac{\widehat{p}_j}{h} I(x \in B_j). \tag{6.7}
$$

To understand the motivation for this estimator, note that, for $x \in B_j$ and h small,

$$\mathbb{E}(\widehat{f}_n(x)) = \frac{\mathbb{E}(\widehat{p}_j)}{h} = \frac{p_j}{h} = \frac{\int_{B_j} f(u)du}{h} \approx \frac{f(x)h}{h} = f(x).$$

6.8 Example. Figure 6.2 shows three different histograms based on $n = 1,266$ data points from an astronomical sky survey. These are the data from Example 4.3. Each data point represents a "redshift," roughly speaking, the distance from us to a galaxy. Choosing the right number of bins involves finding a good tradeoff between bias and variance. We shall see later that the top left histogram has too many bins resulting in oversmoothing and too much bias. The bottom left histogram has too few bins resulting in undersmoothing. The top right histogram is based on 308 bins (chosen by cross-validation). The histogram reveals the presence of clusters of galaxies. ■

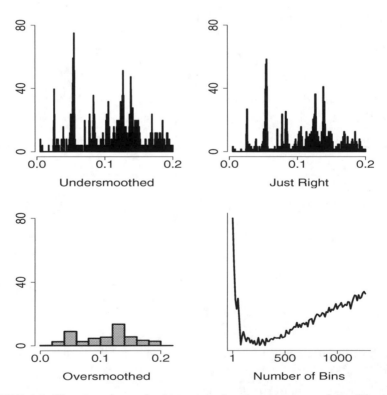

FIGURE 6.2. Three versions of a histogram for the astronomy data. The top left histogram has too many bins. The bottom left histogram has too few bins. The top right histogram uses 308 bins (chosen by cross-validation). The lower right plot shows the estimated risk versus the number of bins.

6.9 Theorem. *Consider fixed x and fixed m, and let B_j be the bin containing x. Then,*

$$\mathbb{E}(\widehat{f}_n(x)) = \frac{p_j}{h} \quad \text{and} \quad \mathbb{V}(\widehat{f}_n(x)) = \frac{p_j(1 - p_j)}{nh^2}. \tag{6.10}$$

6.11 Theorem. *Suppose that f' is absolutely continuous and that $\int (f'(u))^2 du < \infty$. Then*

$$R(\widehat{f}_n, f) = \frac{h^2}{12} \int (f'(u))^2 du + \frac{1}{nh} + o(h^2) + o\left(\frac{1}{n}\right). \tag{6.12}$$

The value h^* that minimizes (6.12) is

$$h^* = \frac{1}{n^{1/3}} \left(\frac{6}{\int (f'(u))^2 du} \right)^{1/3}. \tag{6.13}$$

With this choice of binwidth,

$$R(\widehat{f}_n, f) \sim \frac{C}{n^{2/3}} \tag{6.14}$$

where $C = (3/4)^{2/3} \left(\int (f'(u))^2 du \right)^{1/3}$.

The proof of Theorem 6.11 is in the appendix. We see that with an optimally chosen binwidth, the risk decreases to 0 at rate $n^{-2/3}$. We will see shortly that kernel estimators converge at the faster rate $n^{-4/5}$ and that, in a certain sense, no faster rate is possible; see Theorem 6.31. The formula for the optimal binwidth h^* is of theoretical interest but it is not useful in practice since it depends on the unknown function f. In practice, we use cross-validation as described in Section 6.1. There is a simple formula for computing the cross-validation score $\widehat{J}(h)$.

6.15 Theorem. *The following identity holds:*

$$\widehat{J}(h) = \frac{2}{h(n-1)} - \frac{n+1}{h(n-1)} \sum_{j=1}^{m} \widehat{p}_j^2. \tag{6.16}$$

6.17 Example. We used cross-validation in the astronomy example. We find that $m = 308$ is an approximate minimizer. The histogram in the top right plot in Figure 6.2 was constructed using $m = 308$ bins. The bottom right plot shows the estimated risk, or more precisely, $\widehat{J}$, plotted versus the number of bins. ∎

Next, we want a confidence set for f. Suppose $\widehat{f}_n$ is a histogram with m bins and binwidth $h = 1/m$. For reasons explained in Section 5.7, it is difficult to construct a confidence set for f. Instead, we shall make confidence statements about f at the resolution of the histogram. Thus, define

$$\overline{f}_n(x) = \mathbb{E}(\widehat{f}_n(x)) = \sum_{j=1}^m \frac{p_j}{h} I(x \in B_j) \tag{6.18}$$

where $p_j = \int_{B_j} f(u)du$. Think of $\overline{f}_n(x)$ as a "histogramized" version of f. Recall that a pair of functions (ℓ, u) is a $1 - \alpha$ confidence band for $\overline{f}_n$ if

$$\mathbb{P}\big(\ell(x) \leq \overline{f}_n(x) \leq u(x) \quad \text{for all } x\big) \geq 1 - \alpha. \tag{6.19}$$

We could use the type of reasoning as in (5.100) but, instead, we take a simpler route.

6.20 Theorem. *Let $m = m(n)$ be the number of bins in the histogram $\widehat{f}_n$. Assume that $m(n) \to \infty$ and $m(n) \log n/n \to 0$ as $n \to \infty$. Define*

$$\ell_n(x) = \left(\max\left\{\sqrt{\widehat{f}_n(x)} - c, 0\right\}\right)^2$$

$$u_n(x) = \left(\sqrt{\widehat{f}_n(x)} + c\right)^2 \tag{6.21}$$

where

$$c = \frac{z_{\alpha/(2m)}}{2}\sqrt{\frac{m}{n}}. \tag{6.22}$$

Then, $(\ell_n(x), u_n(x))$ is an approximate $1 - \alpha$ confidence band for $\overline{f}_n$.

PROOF. Here is an outline of the proof. From the central limit theorem, and assuming $1 - p_j \approx 1$, $\widehat{p}_j \approx N(p_j, p_j(1 - p_j)/n)$. By the delta method, $\sqrt{\widehat{p}_j} \approx N(\sqrt{p_j}, 1/(4n))$. Moreover, the $\sqrt{\widehat{p}_j}$'s are approximately independent. Therefore,

$$2\sqrt{n}\left(\sqrt{\widehat{p}_j} - \sqrt{p_j}\right) \approx Z_j \tag{6.23}$$

where $Z_1, \ldots, Z_m \sim N(0, 1)$. Let

$$A = \left\{\ell_n(x) \leq \overline{f}_n(x) \leq u_n(x) \text{ for all } x\right\} = \left\{\max_x \left|\sqrt{\widehat{f}_n(x)} - \sqrt{\overline{f}(x)}\right| \leq c\right\}.$$

Then,

$$\mathbb{P}(A^c) = \mathbb{P}\left(\max_x \left|\sqrt{\widehat{f}_n(x)} - \sqrt{\overline{f}(x)}\right| > c\right)$$

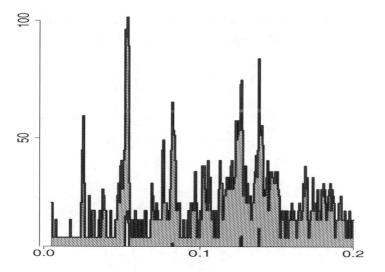

FIGURE 6.3. Ninety-five percent confidence envelope for astronomy data using $m = 308$ bins.

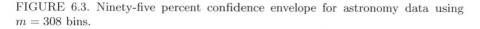

$$= \quad \mathbb{P}\left(\max_j 2\sqrt{n} \left| \sqrt{\widehat{p}_j} - \sqrt{p_j} \right| > z_{\alpha/(2m)} \right)$$

$$\approx \quad \mathbb{P}\left(\max_j |Z_j| > z_{\alpha/(2m)} \right) \leq \sum_{j=1}^{m} \mathbb{P}\left(|Z_j| > z_{\alpha/(2m)} \right)$$

$$= \quad \sum_{j=1}^{m} \frac{\alpha}{m} = \alpha. \quad \blacksquare$$

6.24 Example. Figure 6.3 shows a 95 percent confidence envelope for the astronomy data. We see that even with over 1000 data points, there is still substantial uncertainty about f as reflected by the wide bands. $\blacksquare$

6.3   Kernel Density Estimation

Histograms are not smooth. In this section we discuss kernel density estimators which are smoother and which converge to the true density faster. Recall that the word **kernel** refers to any smooth function K satisfying the conditions given in (4.22). See Section 4.2 for examples of kernels.

6.25 Definition. *Given a kernel K and a positive number h, called the* **bandwidth,** *the* **kernel density estimator** *is defined to be*

$$\widehat{f}_n(x) = \frac{1}{n}\sum_{i=1}^{n}\frac{1}{h}K\left(\frac{x-X_i}{h}\right). \tag{6.26}$$

This amounts to placing a smoothed out lump of mass of size $1/n$ over each data point X_i; see Figure 6.4.

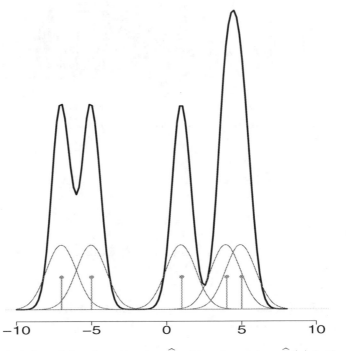

FIGURE 6.4. A kernel density estimator $\widehat{f}_n$. At each point x, $\widehat{f}_n(x)$ is the average of the kernels centered over the data points X_i. The data points are indicated by short vertical bars. The kernels are not drawn to scale.

As with kernel regression, the choice of kernel K is not crucial, but the choice of bandwidth h is important. Figure 6.5 shows density estimates with several different bandwidths. (This is the same as Figure 4.3.) Look also at Figure 6.1. We see how sensitive the estimate $\widehat{f}_n$ is to the choice of h. Small bandwidths give very rough estimates while larger bandwidths give smoother estimates. In general we will let the bandwidth depend on the sample size so we write h_n. Here are some properties of $\widehat{f}_n$.

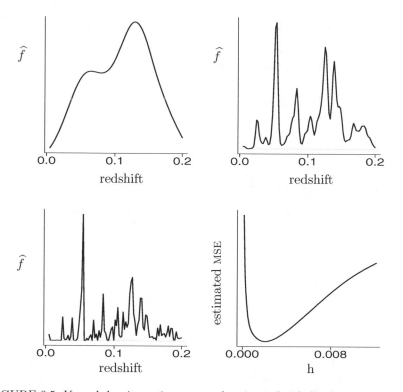

FIGURE 6.5. Kernel density estimators and estimated risk for the astronomy data. Top left: oversmoothed. Top right: just right (bandwidth chosen by cross-validation). Bottom left: undersmoothed. Bottom right: cross-validation curve as a function of bandwidth h. The bandwidth was chosen to be the value of h where the curve is a minimum.

6.27 Theorem. *Assume that f is continuous at x and that $h_n \to 0$ and $nh_n \to \infty$ as $n \to \infty$. Then $\widehat{f}_n(x) \xrightarrow{\text{P}} f(x)$.*

6.28 Theorem. *Let $R_x = \mathbb{E}(f(x) - \widehat{f}(x))^2$ be the risk at a point x and let $R = \int R_x \, dx$ denote the integrated risk. Assume that f'' is absolutely continuous and that $\int (f'''(x))^2 dx < \infty$. Also, assume that K satisfies (4.22). Then,*

$$R_x = \frac{1}{4}\sigma_K^4 h_n^4 (f''(x))^2 + \frac{f(x)\int K^2(x)dx}{nh_n} + O\left(\frac{1}{n}\right) + O(h_n^6)$$

and

$$R = \frac{1}{4}\sigma_K^4 h_n^4 \int (f''(x))^2 dx + \frac{\int K^2(x)dx}{nh} + O\left(\frac{1}{n}\right) + O(h_n^6) \qquad (6.29)$$

where $\sigma_K^2 = \int x^2 K(x)dx$.

PROOF. Write $K_h(x, X) = h^{-1}K((x - X)/h)$ and $\widehat{f}_n(x) = n^{-1}\sum_i K_h(x, X_i)$. Thus, $\mathbb{E}[\widehat{f}_n(x)] = \mathbb{E}[K_h(x, X)]$ and $\mathbb{V}[\widehat{f}_n(x)] = n^{-1}\mathbb{V}[K_h(x, X)]$. Now,

$$
\begin{aligned}
\mathbb{E}[K_h(x, X)] &= \int \frac{1}{h}K\left(\frac{x - t}{h}\right) f(t)\, dt \\
&= \int K(u)f(x - hu)\, du \\
&= \int K(u)\left[f(x) - huf'(x) + \frac{h^2 u^2}{2}f''(x) + \cdots\right] du \\
&= f(x) + \frac{1}{2}h^2 f''(x)\int u^2 K(u)\, du \cdots
\end{aligned}
$$

since $\int K(x)\, dx = 1$ and $\int x\, K(x)\, dx = 0$. The bias is

$$
\mathbb{E}[K_{h_n}(x, X)] - f(x) = \frac{1}{2}\sigma_K^2 h_n^2 f''(x) + O(h_n^4).
$$

By a similar calculation,

$$
\mathbb{V}[\widehat{f}_n(x)] = \frac{f(x)\int K^2(x)\, dx}{n\, h_n} + O\left(\frac{1}{n}\right).
$$

The first result then follows since the risk is the squared bias plus variance. The second result follows from integrating the first. ∎

If we differentiate (6.29) with respect to h and set it equal to 0, we see that the asymptotically optimal bandwidth is

$$
h_* = \left(\frac{c_2}{c_1^2 A(f)n}\right)^{1/5} \tag{6.30}
$$

where $c_1 = \int x^2 K(x)dx$, $c_2 = \int K(x)^2 dx$ and $A(f) = \int (f''(x))^2 dx$. This is informative because it tells us that the best bandwidth decreases at rate $n^{-1/5}$. Plugging h_* into (6.29), we see that if the optimal bandwidth is used then $R = O(n^{-4/5})$. As we saw, histograms converge at rate $O(n^{-2/3})$ showing that kernel estimators are superior in rate to histograms. According to the next theorem, there does not exist an estimator that converges faster than $O(n^{-4/5})$. For a proof, see, for example, Chapter 24 of van der Vaart (1998).

6.31 Theorem. *Let $\mathcal{F}$ be the set of all probability density functions and let $f^{(m)}$ denote the mth derivative of f. Define*

$$
\mathcal{F}_m(c) = \left\{f \in \mathcal{F} : \int |f^{(m)}(x)|^2 dx \le c^2\right\}.
$$

For any estimator $\widehat{f}_n$,

$$\sup_{f \in \mathcal{F}_m(c)} \mathbb{E}_f \int (\widehat{f}_n(x) - f(x))^2 dx \geq b \left(\frac{1}{n}\right)^{2m/(2m+1)} \tag{6.32}$$

where $b > 0$ is a universal constant that depends only on m and c.

In particular, taking $m = 2$ in the previous theorem we see that $n^{-4/5}$ is the fastest possible rate.

In practice, the bandwidth can be chosen by cross-validation but first we describe another method which is sometimes used when f is thought to be very smooth. Specifically, we compute h_* from (6.30) under the idealized assumption that f is Normal. This yields $h_* = 1.06\sigma n^{-1/5}$. Usually, σ is estimated by $\min\{s, Q/1.34\}$ where s is the sample standard deviation and Q is the interquartile range.[1] This choice of h_* works well if the true density is very smooth and is called the **Normal reference rule**.

The Normal Reference Rule

For smooth densities and a Normal kernel, use the bandwidth

$$h_n = \frac{1.06\,\widehat{\sigma}}{n^{1/5}}$$

where

$$\widehat{\sigma} = \min\left\{s, \frac{Q}{1.34}\right\}.$$

Since we don't want to necessarily assume that f is very smooth, it is usually better to estimate h using cross-validation. Recall from Section 6.1 that the cross-validation score is

$$\widehat{J}(h) = \int \widehat{f}^2(x)dx - \frac{2}{n}\sum_{i=1}^{n} \widehat{f}_{-i}(X_i) \tag{6.33}$$

where $\widehat{f}_{-i}$ denotes the kernel estimator obtained by omitting X_i. The next theorem gives a simpler expression for $\widehat{J}$.

6.34 Theorem. *For any $h > 0$,*

$$\mathbb{E}\left[\widehat{J}(h)\right] = \mathbb{E}\left[J(h)\right].$$

---

[1] Recall that the interquartile range is the 75th percentile minus the 25th percentile. The reason for dividing by 1.34 is that $Q/1.34$ is a consistent estimate of σ if the data are from a $N(\mu, \sigma^2)$.

Also,

$$\widehat{J}(h) = \frac{1}{hn^2} \sum_i \sum_j K^* \left(\frac{X_i - X_j}{h} \right) + \frac{2}{nh} K(0) + O \left(\frac{1}{n^2} \right) \qquad (6.35)$$

where $K^*(x) = K^{(2)}(x) - 2K(x)$ *and* $K^{(2)}(z) = \int K(z-y)K(y)dy$.

6.36 Remark. When K is a N(0,1) Gaussian kernel then $K^{(2)}(z)$ is the $N(0,2)$ density. Also, we should point out that the estimator $\widehat{f}_n$ and the cross-validation score (6.35) can be computed quickly using the fast Fourier transform; see pages 61–66 of Silverman (1986).

A justification for cross-validation is given by the following remarkable theorem due to Stone (1984).

6.37 Theorem (Stone's theorem). *Suppose that f is bounded. Let $\widehat{f}_h$ denote the kernel estimator with bandwidth h and let $\widehat{h}$ denote the bandwidth chosen by cross-validation. Then,*

$$\frac{\int \left(f(x) - \widehat{f}_{\widehat{h}}(x) \right)^2 dx}{\inf_h \int \left(f(x) - \widehat{f}_h(x) \right)^2 dx} \xrightarrow{\text{a.s.}} 1. \qquad (6.38)$$

The bandwidth for the density estimator in the upper right panel of Figure 6.5 is based on cross-validation. In this case it worked well but of course there are lots of examples where there are problems. Do not assume that, if the estimator $\widehat{f}$ is wiggly, then cross-validation has let you down. The eye is not a good judge of risk.

Another approach to bandwidth selection called **plug-in bandwidths.** The idea is as follows. The (asymptotically) optimal bandwidth is given in equation (6.30). The only unknown quantity in that formula is $A(f) = \int (f''(x))^2 dx$. If we have an estimate $\widehat{f}''$ of f'', then we can plug this estimate into the formula for the optimal bandwidth h_*. There is a rich and interesting literature on this and similar approaches. The problem with this approach is that estimating f'' is harder than estimating f. Indeed, we need to make stronger assumptions about f to estimate f''. But if we make these stronger assumptions then the (usual) kernel estimator for f is not appropriate. Loader (1999b) has investigated this issue in detail and provides evidence that the plug-in bandwidth approach might not be reliable. There are also methods that apply corrections to plug-in rules; see Hjort (1999).

A generalization of the kernel method is to use **adaptive kernels** where one uses a different bandwidth $h(x)$ for each point x. One can also use a different bandwidth $h(x_i)$ for each data point. This makes the estimator more flexible and allows it to adapt to regions of varying smoothness. But now we have the very difficult task of choosing many bandwidths instead of just one. See Chapter 9 for more on adaptive methods.

Constructing confidence bands for kernel density estimators is more complicated than for regression. We discuss one possible approach in Section 6.6.

6.4   Local Polynomials

In Chapter 5 we saw that kernel regression suffers from boundary bias and that this bias can be reduced by using local polynomials. The same is true for kernel density estimation. But what density estimation method corresponds to local polynomial regression? One possibility, developed by Loader (1999a) and Hjort and Jones (1996) is to use local likelihood density estimation.

The usual definition of log-likelihood is $\mathcal{L}(f) = \sum_{i=1}^{n} \log f(X_i)$. It is convenient to generalize this definition to

$$\mathcal{L}(f) = \sum_{i=1}^{n} \log f(X_i) - n\left(\int f(u)\,du - 1\right).$$

The second term is zero when f integrates to one. Including this term allows us to maximize over all non-negative f with imposing the constraint that $\int f(u)du = 1$. The local version of this log-likelihood is as follows:

6.39 Definition. *Given a kernel K and bandwidth h, the local log-likelihood at a target value x is*

$$\mathcal{L}_x(f) = \sum_{i=1}^{n} K\left(\frac{X_i - x}{h}\right) \log f(X_i) - n \int K\left(\frac{u - x}{h}\right) f(u)\,du. \quad (6.40)$$

The above definition is for an arbitrary density f. We are interested in the case where we approximate $\log f(u)$ by a polynomial in a neighborhood of x. Thus we write

$$\log f(u) \approx P_x(a, u) \qquad (6.41)$$

where

$$P_x(a, u) = a_0 + a_1(x - u) + \cdots + a_p \frac{(x - u)^p}{p!}. \qquad (6.42)$$

Plugging (6.41) into (6.40) yields the **local polynomial log-likelihood**

$$\mathcal{L}_x(a) = \sum_{i=1}^{n} K\left(\frac{X_i - x}{h}\right) P_x(a, X_i) - n \int K\left(\frac{u - x}{h}\right) e^{P_x(a,u)} du. \quad (6.43)$$

6.44 Definition. *Let* $\widehat{a} = (\widehat{a}_0, \ldots, \widehat{a}_p)^T$ *maximize* $\mathcal{L}_x(a)$. *The* **local likelihood density estimate** *is*

$$\widehat{f}_n(x) = e^{P_x(\widehat{a}, x)} = e^{\widehat{a}_0}. \quad (6.45)$$

6.46 Remark. When $p = 0$, $\widehat{f}_n$ reduces to kernel density estimation.

6.5   Multivariate Problems

Suppose now that the data are d-dimensional so that $X_i = (X_{i1}, \ldots, X_{id})$. As we discussed in the two previous chapters, we can generalize the methods to higher dimensions quite easily in principle, though the curse of dimensionality implies that the accuracy of the estimator deteriorates quickly as dimension increases.

The kernel estimator can easily be generalized to d dimensions. Most often, we use the product kernel

$$\widehat{f}_n(x) = \frac{1}{nh_1 \cdots h_d} \sum_{i=1}^{n} \left\{ \prod_{j=1}^{d} K\left(\frac{x_j - X_{ij}}{h_j}\right) \right\}. \quad (6.47)$$

The risk is given by

$$R \approx \frac{1}{4}\sigma_K^4 \left[\sum_{j=1}^{d} h_j^4 \int f_{jj}^2(x)dx + \sum_{j \neq k} h_j^2 h_k^2 \int f_{jj} f_{kk} dx \right] + \frac{\left(\int K^2(x)dx\right)^d}{nh_1 \cdots h_d} \quad (6.48)$$

where f_{jj} is the second partial derivative of f. The optimal bandwidth satisfies $h_i = O(n^{-1/(4+d)})$ leading to a risk of order $R = O(n^{-4/(4+d)})$. Again, we see that the risk increases quickly with dimension. To get a sense of how serious this problem is, consider the following table from Silverman (1986) which shows the sample size required to ensure a relative mean squared error less than 0.1 at 0 when the density is multivariate normal and the optimal bandwidth is selected.

Dimension	Sample Size
1	4
2	19
3	67
4	223
5	768
6	2790
7	10,700
8	43,700
9	187,000
10	842,000

This is bad news. If you do attempt to estimate a density in a high dimensional problem, you should not report the results without reporting confidence bands. The confidence band method described in Section 6.6 can be extended for the multivariate case, although we do not report the details here. These bands get very wide as d increases. The problem is not the method of estimation; rather, the wide bands correctly reflect the difficulty of the problem.

6.6   Converting Density Estimation Into Regression

There is a useful trick for converting a density estimation problem into a regression problem. Then we can use all the regression methods from the previous chapter. This trick is an old idea but was recently made rigorous by Nussbaum (1996a) and Brown et al. (2005). By converting to regression, we can use all the tools we developed in the previous chapter, including the method for constructing confidence bands.

Suppose $X_1, \ldots, X_n \sim F$ with density $f = F'$. For simplicity, suppose the data are on $[0,1]$. Divide the interval $[0,1]$ into k equal width bins where $k \approx n/10$. Define

$$Y_j = \sqrt{\frac{k}{n}} \times \sqrt{N_j + \frac{1}{4}} \tag{6.49}$$

where N_j is the number of observations in bin j. Then,

$$Y_j \approx r(t_j) + \sigma \epsilon_j \tag{6.50}$$

where $\epsilon_j \sim N(0,1)$, $\sigma = \sqrt{\frac{k}{4n}}$, $r(x) = \sqrt{f(x)}$ and t_j is the midpoint of the jth bin. To see why, let B_j denote the jth bin and note that

$$N_j \approx \text{Poisson}\left(n \int_{B_j} f(x)dx\right) \approx \text{Poisson}\left(\frac{nf(t_j)}{k}\right)$$

so that $\mathbb{E}(N_j) = V(N_j) \approx nf(t_j)/k$. Applying the delta method, we see that $\mathbb{E}(Y_j) \approx \sqrt{f(t_j)}$ and $\mathbb{V}(Y_j) \approx k/(4n)$.

We have thus converted the density estimation problem into a nonparametric regression problem with equally spaced x_is and constant variance. We can now apply any nonparametric regression method to get an estimate $\widehat{r}_n$ and take

$$\widehat{f}_n(x) = \frac{(r^+(x))^2}{\int_0^1 (r^+(s))^2 \, ds}$$

where $r^+(x) = \max\{\widehat{r}_n(x), 0\}$. In particular, we can construct confidence bands as in Chapter 5. It is important to note that binning is not a smoothing step; binning is being used to turn density estimation into regression.

6.51 Example. Figure 6.6 shows the method applied to data from the Bart Simpson distribution. The top plot shows the cross-validation score. The bottom plot shows the estimated density and 95 percent confidence bands. ∎

6.7   Bibliographic Remarks

Kernel smoothing was invented by Rosenblatt (1956) and Parzen (1962). The cross-validation method is due to Rudemo (1982). Two very good books on density estimation are Scott (1992) and Silverman (1986). For information on a different approach called the **scale-space approach** see Chaudhuri and Marron (1999) and Chaudhuri and Marron (2000).

6.8   Appendix

PROOF OF THEOREM 6.11. For any $x, u \in B_j$,

$$f(u) = f(x) + (u - x)f'(x) + \frac{(u - x)^2}{2} f''(\widetilde{x})$$

for some $\widetilde{x}$ between x and u.

Hence,

$$p_j = \int_{B_j} f(u)du = \int_{B_j} \left(f(x) + (u - x)f'(x) + \frac{(u - x)^2}{2} f''(\widetilde{x}) \right) du$$

$$= f(x)h + hf'(x) \left(h \left(j - \frac{1}{2} \right) - x \right) + O(h^3).$$

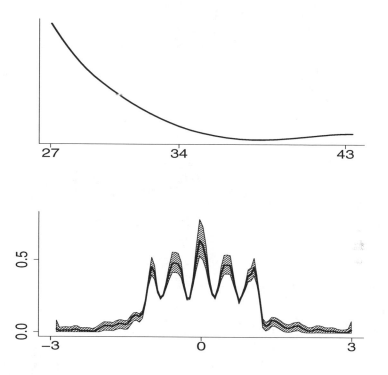

FIGURE 6.6. Density estimation by regression. The data were binned and kernel regression was used on the square root of the counts. The top plot shows the cross-validation score by effective degrees of freedom. The bottom plot shows the estimate and a 95 percent confidence envelope. See Example 6.51.

Therefore, the bias $b(x)$ is

$$
\begin{aligned}
b(x) &= \mathbb{E}(\widehat{f}_n(x)) - f(x) = \frac{p_j}{h} - f(x) \\
&= \frac{f(x)h + hf'(x)\left(h\left(j - \frac{1}{2}\right) - x\right) + O(h^3)}{h} - f(x) \\
&= f'(x)\left(h\left(j - \frac{1}{2}\right) - x\right) + O(h^2).
\end{aligned}
$$

By the mean value theorem we have, for some $\widetilde{x}_j \in B_j$, that

$$
\begin{aligned}
\int_{B_j} b^2(x)\, dx &= \int_{B_j} (f'(x))^2 \left(h\left(j - \frac{1}{2}\right) - x\right)^2 dx + O(h^4) \\
&= (f'(\widetilde{x}_j))^2 \int_{B_j} \left(h\left(j - \frac{1}{2}\right) - x\right)^2 dx + O(h^4)
\end{aligned}
$$

$$= \quad (f'(\widetilde{x}_j))^2 \frac{h^3}{12} + O(h^4).$$

Therefore,

$$
\begin{aligned}
\int_0^1 b^2(x)dx &= \sum_{j=1}^m \int_{B_j} b^2(x)dx + O(h^3) \\
&= \sum_{j=1}^m (f'(\widetilde{x}_j))^2 \frac{h^3}{12} + O(h^3) \\
&= \frac{h^2}{12} \sum_{j=1}^m h\, (f'(\widetilde{x}_j))^2 + O(h^3) \\
&= \frac{h^2}{12} \int_0^1 (f'(x))^2 dx + o(h^2).
\end{aligned}
$$

Now consider the variance. By the mean value theorem, $p_j = \int_{B_j} f(x)dx = hf(x_j)$ for some $x_j \in B_j$. Hence, with $v(x) = \mathbb{V}(\widehat{f}_n(x))$,

$$
\begin{aligned}
\int_0^1 v(x)\, dx &= \sum_{j=1}^m \int_{B_j} v(x)\, dx = \sum_{j=1}^m \int_{B_j} \frac{p_j(1-p_j)}{nh^2} \\
&= \frac{1}{nh^2} \sum_{j=1}^m \int_{B_j} p_j - \frac{1}{nh^2} \sum_{j=1}^m \int_{B_j} p_j^2 = \frac{1}{nh} - \frac{1}{nh} \sum_{j=1}^m p_j^2 \\
&= \frac{1}{nh} - \frac{1}{nh} \sum_{j=1}^m h^2 f^2(x_j) = \frac{1}{nh} - \frac{1}{n} \sum_{j=1}^m h f^2(x_j) \\
&= \frac{1}{nh} - \frac{1}{n} \left(\int_0^1 f^2(x)dx + o(1) \right) = \frac{1}{nh} + o\left(\frac{1}{n}\right). \quad \blacksquare
\end{aligned}
$$

6.9   Exercises

1. Prove Theorem 6.27.

2. Let $X_1, \ldots, X_n \sim f$ and let $\widehat{f}_n$ be the kernel density estimator using the boxcar kernel:
$$
K(x) = \begin{cases} 1 & -\frac{1}{2} < x < \frac{1}{2} \\ 0 & \text{otherwise.} \end{cases}
$$

(a) Show that
$$
\mathbb{E}(\widehat{f}(x)) = \frac{1}{h} \int_{x-(h/2)}^{x+(h/2)} f(y)dy
$$

and

$$\mathbb{V}(\widehat{f}(x)) = \frac{1}{nh^2} \left[\int_{x-(h/2)}^{x+(h/2)} f(y)dy - \left(\int_{x-(h/2)}^{x+(h/2)} f(y)dy \right)^2 \right].$$

(b) Show that if $h \to 0$ and $nh \to \infty$ as $n \to \infty$ then $\widehat{f}_n(x) \overset{P}{\longrightarrow} f(x)$.

3. Prove that $\widehat{J}(h)$ is an unbiased estimate of $J(h)$ for histograms and kernel density estimates.

4. Prove equation 6.35.

5. Data on the salaries of the chief executive officer of 60 companies is available at:

   http://lib.stat.cmu.edu/DASL/Datafiles/ceodat.html

   Investigate the distribution of salaries using a histogram and a kernel density estimator. Use least squares cross-validation to choose the amount of smoothing. Also consider the Normal reference rule for picking a bandwidth for the kernel. There appear to be a few bumps in the density. Are they real? Use confidence bands to address this question. Finally, try using various kernel shapes and comment on the resulting estimates.

6. Get the data on fragments of glass collected in forensic work from the book website. Estimate the density of the first variable (refractive index) using a histogram and a kernel density estimator. Use cross-validation to choose the amount of smoothing. Experiment with different binwidths and bandwidths. Comment on the similarities and differences. Construct 95 percent confidence bands for your estimators. For the kernel, experiment with different kernel shapes.

7. Consider the data in Exercise 6. Examine the fit as a function of the bandwidth h. Do this by plotting the fit for many values of h. Add confidence bands to all the fits. If you are feeling very ambitious, read Chaudhuri and Marron (1999) and apply that method.

8. Prove that local likelihood density estimation reduces to kernel density estimation when the degree of the polynomial $p = 0$.

9. Apply local polynomial density estimation to the data in Exercise 6.

10. Generate data from the Bart Simpson distribution (6.2). Compare kernel density estimation to the method in Section 6.6. Try the following sample sizes: $n = 25, 50, 100, 1,000$.

7
Normal Means and Minimax Theory

In this chapter we will discuss the **many Normal means problem** which unifies some nonparametric problems and will be the basis for the methods in the next two chapters. The material in this chapter is more theoretical than in the rest of the book. If you are not interested in the theoretical details, I recommend reading sections 7.1, 7.2, and 7.3 and then skipping to the next chapter, referring back as needed. If you want more details on this topic, I recommend Johnstone (2003).

7.1 The Normal Means Model

Let $Z^n = (Z_1, \ldots, Z_n)$ where

$$Z_i = \theta_i + \sigma_n \, \epsilon_i, \quad i = 1, \ldots, n, \tag{7.1}$$

$\epsilon_1, \ldots, \epsilon_n$ are independent, Normal$(0, 1)$ random variables,

$$\theta^n = (\theta_1, \ldots, \theta_n) \in \mathbb{R}^n$$

is a vector of unknown parameters and σ_n is assumed known. Typically $\sigma_n = \sigma/\sqrt{n}$ but we shall not assume this unless specifically noted. Sometimes we write Z^n and θ^n as Z and θ. The model may appear to be parametric but the number of parameters is increasing at the same rate as the number of

θ_1	θ_2	$\ldots$	θ_i	$\ldots$	θ_n
X_{11}	X_{21}	$\ldots$	X_{i1}	$\ldots$	X_{n1}
$\vdots$	$\vdots$	$\vdots$	$\vdots$	$\vdots$	$\vdots$
X_{1j}	X_{2j}	$\ldots$	X_{ij}	$\ldots$	X_{nj}
$\vdots$	$\vdots$	$\vdots$	$\vdots$	$\vdots$	$\vdots$
X_{1n}	X_{2n}	$\ldots$	X_{in}	$\vdots$	X_{nn}
Z_1	Z_2	$\ldots$	Z_i	$\ldots$	Z_n

FIGURE 7.1. The Normal means model. $X_{ij} = \theta_i + N(0, \sigma^2)$ and $Z_i = n^{-1} \sum_{j=1}^{n} X_{ij} = \theta_i + \sigma_n \epsilon_i$ where $\sigma_n = \sigma/\sqrt{n}$. Estimating the parameters $\theta_1, \ldots, \theta_n$ from the n column means $Z_1, \ldots, Z_n$ leads to the model (7.1) with $\sigma_n = \sigma/\sqrt{n}$.

data points. This model carries with it all the complexities and subtleties of a nonparametric problem. We will also consider an infinite-dimensional version of the model:

$$Z_i = \theta_i + \sigma_n \, \epsilon_i, \quad i = 1, 2, \ldots, \tag{7.2}$$

where now the unknown parameter is $\theta = (\theta_1, \theta_2, \ldots)$.

Throughout this chapter we take σ_n^2 as known. In practice, we would need to estimate the variance using one of the methods discussed in Chapter 5. In this case, the exact results that follow may no longer hold but, under appropriate smoothness conditions, asymptotic versions of the results will hold.

7.3 Example. To provide some intuition for this model, suppose that we have data $X_{ij} = \theta_i + \sigma \delta_{ij}$ where $1 \leq i, j \leq n$ and the δ_{ij} are independent N(0,1) random variables. This is simply a one-way analysis of variance model; see Figure 7.1. Let $Z_i = n^{-1} \sum_{j=1}^{n} X_{ij}$. Then the model (7.1) holds with $\sigma_n = \sigma/\sqrt{n}$. We get the infinite version (7.2) by having infinitely many columns in Figure 7.1 (but still n rows). ∎

Given an estimator $\widehat{\theta}^n = (\widehat{\theta}_1, \ldots, \widehat{\theta}_n)$ we will use the squared error loss

$$L(\widehat{\theta}^n, \theta^n) = \sum_{i=1}^{n} (\widehat{\theta}_i - \theta_i)^2 = ||\widehat{\theta}^n - \theta^n||^2$$

with risk function

$$R(\widehat{\theta}^n, \theta^n) = \mathbb{E}_\theta \big(L(\widehat{\theta}^n, \theta^n) \big) = \sum_{i=1}^{n} \mathbb{E}_\theta (\widehat{\theta}_i - \theta_i)^2.$$

An obvious choice for an estimator of θ^n is $\widehat{\theta}^n = Z^n$. This estimator has impressive credentials: it is the maximum likelihood estimator, it is the minimum variance unbiased estimator and it is the Bayes estimator under a flat prior. Nonetheless, it is a poor estimator. Its risk is

$$R(Z^n, \theta^n) = \sum_{i=1}^{n} \mathbb{E}_\theta (Z_i - \theta_i)^2 = \sum_{i=1}^{n} \sigma_n^2 = n\sigma_n^2.$$

We shall see that there are estimators with substantially smaller risk.

Before we explain how we can improve on the MLE, let's first see how the normal means problem relates to nonparametric regression and density estimation. To do that, we need to review some theory about function spaces.

7.2   Function Spaces

Let $L_2(a, b)$ denote the set of functions $f : [a, b] \to \mathbb{R}$ such that $\int_a^b f^2(x)\, dx < \infty$. Unless otherwise indicated, assume that $a = 0$ and $b = 1$. The **inner product** between two functions f and g in $L_2(a, b)$ is $\int_a^b f(x)g(x)dx$ and the **norm** of f is $||f|| = \sqrt{\int_a^b f^2(x)\, dx}$. A sequence of functions $\phi_1, \phi_2, \ldots$ is called **orthonormal** if $||\phi_j|| = 1$ for all j (normalized) and $\int_a^b \phi_i(x)\phi_j(x)dx = 0$ for $i \neq j$ (orthogonal). The sequence is **complete** if the only function that is orthogonal to each ϕ_j is the zero function. A complete, orthonormal set of functions forms a **basis**, meaning that if $f \in L_2(a, b)$ then f can be expanded in the basis:

7.4 Theorem. *If $f \in L_2(a, b)$ then*[1]

$$f(x) = \sum_{j=1}^{\infty} \theta_j \phi_j(x) \tag{7.5}$$

where

$$\theta_j = \int_a^b f(x)\phi_j(x)\, dx. \tag{7.6}$$

Furthermore,

$$\int_a^b f^2(x)dx = \sum_{j=1}^{\infty} \theta_j^2 \tag{7.7}$$

which is known as **Parseval's identity.**

---

[1] The equality sign in (7.5) means that $\int_a^b (f(x) - f_N(x))^2 dx \to 0$ as $N \to \infty$, where $f_N = \sum_{j=1}^{N} \theta_j \phi_j(x)$.

An example of an orthonormal basis for $L_2(0,1)$ is the **cosine basis**

$$\phi_0(x) = 1, \qquad \phi_j(x) = \sqrt{2}\cos(2\pi j x), \qquad j = 1, 2, \dots.$$

Another example is the **Legendre basis** defined on $(-1,1)$:

$$P_0(x) = 1, \quad P_1(x) = x, \quad P_2(x) = \frac{1}{2}(3x^2 - 1), \quad P_3(x) = \frac{1}{2}(5x^3 - 3x), \quad \dots$$

These polynomials are defined by the relation

$$P_n(x) = \frac{1}{2^n n!} \frac{d^n}{dx^n} (x^2 - 1)^n.$$

The Legendre polynomials are orthogonal but not orthonormal since

$$\int_{-1}^{1} P_n^2(x) dx = \frac{2}{2n+1}.$$

However, we can define modified Legendre polynomials $Q_n(x) = \sqrt{(2n+1)/2} \, P_n(x)$ which then form an orthonormal basis for $L_2(-1,1)$.

Next we introduce Sobolev spaces, which are sets of smooth functions. Let $D^j f$ denote the jth weak derivative[2] of f.

7.8 Definition. *The* **Sobolev space of order** m, *is defined by*

$$W(m) = \left\{ f \in L_2(0,1) : \; D^m f \in L_2(0,1) \right\}.$$

The **Sobolev space of order** m **and radius** c, *is defined by*

$$W(m,c) = \left\{ f : \; f \in W(m), \; \|D^m f\|^2 \le c^2 \right\}.$$

The **periodic Sobolev class** *is*

$$\widetilde{W}(m,c) = \left\{ f \in W(m,c) : \; D^j f(0) = D^j f(1), \; j = 0, \dots, m-1 \right\}.$$

An **ellipsoid** is a set of the form

$$\Theta = \left\{ \theta : \; \sum_{j=1}^{\infty} a_j^2 \theta_j^2 \le c^2 \right\} \tag{7.9}$$

where a_j is a sequence of numbers such that $a_j \to \infty$ as $j \to \infty$.

---

[2] The weak derivative is defined in the appendix.

7.10 Definition. *If Θ is an ellipsoid and if $a_j^2 \sim (\pi j)^{2m}$ as $j \to \infty$, we call Θ a **Sobolev ellipsoid** or a **Sobolev body** and we denote it by $\Theta(m, c)$.*

Now we relate Sobolev spaces to Sobolev ellipsoids.

7.11 Theorem. *Let $\{\phi_j, j = 0, 1, \ldots\}$ be the* **Fourier basis***:*

$$\phi_1(x) = 1, \quad \phi_{2j}(x) = \frac{1}{\sqrt{2}} \cos(2j\pi x), \quad \phi_{2j+1}(x) = \frac{1}{\sqrt{2}} \sin(2j\pi x), \quad j = 1, 2, \ldots$$

Then,

$$\widetilde{W}(m, c) = \left\{ f : f = \sum_{j=1}^{\infty} \theta_j \phi_j, \ \sum_{j=1}^{\infty} a_j^2 \theta_j^2 \leq c^2 \right\} \tag{7.12}$$

where $a_j = (\pi j)^m$ for j even and $a_j = (\pi(j-1))^m$ for j odd.

Thus, a Sobolev space corresponds to a Sobolev ellipsoid with $a_j \sim (\pi j)^{2m}$. It is also possible to relate the class $W(m, c)$ to an ellipsoid although the details are more complicated; see Nussbaum (1985).

In Sobolev spaces, smooth functions have small coefficients θ_j when j is large, otherwise $\sum_j \theta_j^2 (\pi j)^{2m}$ will blow up. Thus, to smooth a function, we shrink the θ_js closer to zero. Hence:

> smoothing f corresponds to shrinking the θ_j's towards zero for large j.

A generalization of Sobolev spaces are **Besov spaces**. These include Sobolev spaces as a special case but they also include functions with less smoothness. We defer discussion of Besov spaces until Chapter 9.

7.3   Connection to Regression and Density Estimation

Consider the nonparametric regression model

$$Y_i = f(i/n) + \sigma \epsilon_i, \quad i = 1, \ldots, n \tag{7.13}$$

where $\epsilon_i \sim N(0, 1)$, σ is known and $f \in L_2(0, 1)$. Let $\phi_1, \phi_2, \ldots$ be an orthonormal basis and write $f(x) = \sum_{i=1}^{\infty} \theta_j \phi_j(x)$ where $\theta_j = \int f(x) \phi_j(x) dx$. First, approximate f by the finite series $f(x) \approx \sum_{i=1}^{n} \theta_j \phi_j(x)$. Now define

$$Z_j = \frac{1}{n} \sum_{i=1}^{n} Y_i \, \phi_j(i/n) \tag{7.14}$$

for $j = 1, \ldots, n$. The random variable Z_j has a Normal distribution since Z_j is a linear combination of Normals. The mean of Z_j is

$$
\begin{aligned}
\mathbb{E}(Z_j) &= \frac{1}{n} \sum_{i=1}^{n} \mathbb{E}(Y_i) \phi_j(i/n) = \frac{1}{n} \sum_{i=1}^{n} f(i/n) \phi_j(i/n) \\
&\approx \int f(x) \phi_j(x) \, dx = \theta_j.
\end{aligned}
$$

The variance is

$$
\begin{aligned}
\mathbb{V}(Z_j) &= \frac{1}{n^2} \sum_{i=1}^{n} \mathbb{V}(Y_i) \, \phi_j^2(i/n) = \frac{\sigma^2}{n} \frac{1}{n} \sum_{i=1}^{n} \phi_j^2(i/n) \\
&\approx \frac{\sigma^2}{n} \int \phi_j^2(x) dx = \frac{\sigma^2}{n} \equiv \sigma_n^2.
\end{aligned}
$$

A similar calculation shows that $\mathrm{Cov}(Z_j, Z_k) \approx 0$. We conclude that the Z_j are approximately independent and

$$
Z_j \sim N(\theta_j, \sigma_n^2), \quad \sigma_n^2 = \frac{\sigma^2}{n}. \tag{7.15}
$$

We have thus converted the problem of estimating f into the problem of estimating the means of n Normal random variables as in (7.1) with $\sigma_n^2 = \sigma^2/n$. Also, squared error loss for f corresponds to squared error loss for θ since, by Parseval's identity, if $\widehat{f}_n(x) = \sum_{j=1}^{\infty} \widehat{\theta}_j \phi_j(x)$,

$$
\|\widehat{f}_n - f\|^2 = \int \left(\widehat{f}_n(x) - f(x) \right)^2 dx = \sum_{j=1}^{\infty} (\widehat{\theta}_j - \theta_j)^2 = \|\widehat{\theta} - \theta\|^2 \tag{7.16}
$$

where $\|\theta\| = \sqrt{\sum_j \theta_j^2}$.

It turns out that other nonparametric problems, such as density estimation, can also be connected to the Normal means problem. In the case of density estimation, it is the square root of the density that appears in the white noise problem. In this sense, the many Normal means problem serves as a unifying framework for many nonparametric models. See Nussbaum (1996a), Claeskens and Hjort (2004) and the appendix for more details.

7.4   Stein's Unbiased Risk Estimator (SURE)

Let $\widehat{\theta}$ be an estimate of θ. It will be useful to have an estimate of the risk of $\widehat{\theta}$. In previous chapters we used cross-validation to estimate risk. In the present context there is a more elegant method for risk estimation due to Stein (1981) known as **Stein's unbiased risk estimator** or SURE.

7.17 Theorem (Stein). *Let $Z \sim N_n(\theta, V)$, let $\widehat{\theta} = \widehat{\theta}(Z)$ be an estimate of θ and let $g(Z_1, \ldots, Z_n) = \widehat{\theta} - Z$. Note that g maps $\mathbb{R}^n$ to $\mathbb{R}^n$. Define*

$$\widehat{R}(z) = \mathrm{tr}(V) + 2\mathrm{tr}\,(V\,D) + \sum_i g_i^2(z) \qquad (7.18)$$

where tr *denotes the trace of a matrix,* $g_i = \widehat{\theta}_i - Z_i$ *and the (i, j) component of D is the partial derivative of the ith component of $g(z_1, \ldots, z_n)$ with respect to z_j. If g is weakly differentiable[3] then*

$$\mathbb{E}_\theta(\widehat{R}(Z)) = R(\theta, \widehat{\theta}).$$

If we apply Theorem 7.17 to the model (7.1) we get the following.

The SURE Formula for the Normal Means Model

Let $\widehat{\theta}$ be a weakly differentiable estimator of θ in model (7.1). An unbiased estimate of the risk of $\widehat{\theta}$ is

$$\widehat{R}(z) = n\sigma_n^2 + 2\sigma_n^2 \sum_{i=1}^n D_i + \sum_{i=1}^n g_i^2 \qquad (7.19)$$

where $g(Z_1, \ldots, Z_n) = \widehat{\theta}^n - Z^n$ and $D_i = \partial g(z_1, \ldots, z_n)/\partial z_i$.

PROOF OF THEOREM 7.17. We will prove the case where $V = \sigma^2 I$. If $X \sim N(\mu, \sigma^2)$ then $\mathbb{E}(g(X)(X - \mu)) = \sigma^2 \mathbb{E} g'(X)$. (This is known as Stein's Lemma and it can be proved using integration by parts. See Exercise 4.) Hence, $\sigma^2 \mathbb{E}_\theta D_i = \mathbb{E}_\theta g_i(Z_i - \theta)$ and

$$
\begin{aligned}
\mathbb{E}_\theta(\widehat{R}(Z)) &= n\sigma^2 + 2\sigma^2 \sum_{i=1}^n \mathbb{E}_\theta D_i + \sum_{i=1}^n \mathbb{E}_\theta(\widehat{\theta}_i - Z_i)^2 \\
&= n\sigma^2 + 2\sum_{i=1}^n \mathbb{E}_\theta\,(g_i(Z_i - \theta_i)) + \sum_{i=1}^n \mathbb{E}_\theta(\widehat{\theta}_i - Z_i)^2 \\
&= \sum_{i=1}^n \mathbb{E}_\theta(Z_i - \theta_i)^2 + 2\sum_{i=1}^n \mathbb{E}_\theta\left((\widehat{\theta}_i - Z_i)(Z_i - \theta_i)\right) \\
&\quad + \sum_{i=1}^n \mathbb{E}_\theta(\widehat{\theta}_i - Z_i)^2 \\
&= \sum_{i=1}^n \mathbb{E}_\theta(\widehat{\theta}_i - Z_i + Z_i - \theta_i)^2 = \sum_{i=1}^n \mathbb{E}_\theta(\widehat{\theta}_i - \theta_i)^2 = R(\widehat{\theta}, \theta). \quad \blacksquare
\end{aligned}
$$

---

[3]Weak differentiability is defined in the appendix.

7.20 Example. Let $V = \sigma^2 I$. Consider $\widehat{\theta} = Z$. Then $g(z) = (0, \ldots, 0)$ and $\widehat{R}(Z) = n\sigma^2$. In this case, $\widehat{R}$ is equal to the true risk. Now consider the linear estimator $\widehat{\theta} = bZ = (bZ_1, \ldots, bZ_n)$. Hence, $g(Z) = bZ - Z = (b-1)Z$ and $D_i = b - 1$. Therefore, $\widehat{R}(Z) = (2b-1)n\sigma^2 + (1-b)^2 \sum_{i=1}^{n} Z_i^2$. Next consider the **soft threshold estimator** defined by

$$
\widehat{\theta}_i = \begin{cases} Z_i + \lambda & Z_i < -\lambda \\ 0 & -\lambda \leq Z_i \leq \lambda \\ Z_i - \lambda & Z_i > \lambda \end{cases} \tag{7.21}
$$

where $\lambda > 0$ is a constant. We can write this estimator more succinctly as

$$
\widehat{\theta}_i = \text{sign}(Z_i)(|Z_i| - \lambda)_+.
$$

In Exercise 5 you will show that the SURE formula gives

$$
\widehat{R}(Z) = \sum_{i=1}^{n} \left(\sigma^2 - 2\sigma^2 I(|Z_i| \leq \lambda) + \min(Z_i^2, \lambda^2) \right). \tag{7.22}
$$

Finally, consider the **hard threshold estimator** defined by

$$
\widehat{\theta}_i = \begin{cases} Z_i & |Z_i| > \lambda \\ 0 & |Z_i| \leq \lambda \end{cases} \tag{7.23}
$$

where $\lambda > 0$ is a constant. It is tempting to use SURE but this is inappropriate because this estimator is not weakly differentiable. ∎

7.24 Example (Model selection). For each $S \subset \{1, \ldots, n\}$ define

$$
\widehat{\theta}_S = Z_i I(i \in S). \tag{7.25}
$$

We can think of S as a submodel which says that $Z_i \sim N(\theta_i, \sigma_n^2)$ for $i \in S$ and $Z_i \sim N(0, \sigma_n^2)$ for $i \notin S$. Then $\widehat{\theta}_S$ is the estimator of θ assuming the model S. The true risk of $\widehat{\theta}_S$ is

$$
R(\widehat{\theta}_S, \theta) = \sigma_n^2 |S| + \sum_{i \in S^c} \theta_i^2
$$

where $|S|$ denotes the number of points in S. Replacing θ_i^2 in the risk with its unbiased estimator $Z_i^2 - \sigma_n^2$ yields the risk estimator

$$
\widehat{R}_S = \sigma_n^2 |S| + \sum_{i \in S^c} (Z_i^2 - \sigma_n^2). \tag{7.26}
$$

It is easy to check that this corresponds to the SURE formula. Now let $\mathcal{S}$ be some class of sets where each $S \in \mathcal{S}$ is a subset of $\{1, \ldots, n\}$. Choosing $S \in \mathcal{S}$ to minimize $\widehat{R}_S$ is an example of **model selection**. The special case where

$$
\mathcal{S} = \Big\{ \emptyset, \{1\}, \{1, 2\}, \ldots, \{1, 2, \ldots, n\} \Big\}
$$

is called **nested subset selection**. Taking S to be all subsets of $\{1, \ldots, n\}$ corresponds to **all possible subsets**. For any fixed model S, we expect that $\widehat{R}_S$ will be close to $R(\widehat{\theta}_S, \theta)$. However, this does not guarantee that $\widehat{R}_S$ is close to $R(\widehat{\theta}_S, \theta)$ uniformly over S. See Exercise 10. ∎

7.5 Minimax Risk and Pinsker's Theorem

If Θ_n is a subset of $\mathbb{R}^n$, we define the **minimax risk** over Θ_n by

$$R_n \equiv R(\Theta_n) \equiv \inf_{\widehat{\theta}} \sup_{\theta \in \Theta_n} R(\widehat{\theta}, \theta) \tag{7.27}$$

where the infimum is over all estimators. Two questions we will address are: (i) what is the value of the minimax risk $R(\Theta_n)$? and (ii) can we find an estimator that achieves this risk?

The following theorem[4] gives the exact, limiting form of the minimax risk for the L_2 ball

$$\Theta_n(c) = \left\{ (\theta_1, \ldots, \theta_n) : \sum_{i=1}^n \theta_i^2 \le c^2 \right\}.$$

7.28 Theorem (Pinsker's theorem). *Assume the model (7.1) with $\sigma_n^2 = \sigma^2/n$. For any $c > 0$,*

$$\liminf_{n \to \infty} \inf_{\widehat{\theta}} \sup_{\theta \in \Theta_n(c)} R(\widehat{\theta}, \theta) = \frac{\sigma^2 c^2}{\sigma^2 + c^2}. \tag{7.29}$$

The right-hand side of (7.29) gives an exact expression for the (asymptotic) minimax risk. This expression is strictly smaller than σ^2 which is the risk for the maximum likelihood estimator. Later, we will introduce the James–Stein estimator which asymptotically achieves this risk. The proof of the theorem, which is in the appendix, is a bit technical and may be skipped without loss of continuity. Here is the basic idea behind the proof.

First, we note that the estimator with coordinates $\widehat{\theta}_j = c^2 Z_j/(\sigma^2 + c^2)$ has risk bounded above by $\sigma^2 c^2/(\sigma^2 + c^2)$. Hence,

$$R_n \le \frac{\sigma^2 c^2}{\sigma^2 + c^2}. \tag{7.30}$$

If we could find a prior π on $\Theta_n(c)$ whose posterior mean $\widetilde{\theta}$ also has risk $\sigma^2 c^2/(\sigma^2 + c^2)$ then we could argue that, for any estimator $\widehat{\theta}$, we have

$$\frac{\sigma^2 c^2}{\sigma^2 + c^2} = \int R(\theta, \widetilde{\theta}) d\pi(\theta) \le \int R(\theta, \widehat{\theta}) d\pi(\theta) \le \sup_{\theta \in \Theta_n} R(\theta, \widehat{\theta}) = R_n. \tag{7.31}$$

---

[4]This is a finite-dimensional version of Pinsker's theorem. Theorem 7.32 is the usual version.

Combining (7.30) and (7.31) would yield $R_n = \sigma^2 c^2/(\sigma^2 + c^2)$. The proof is essentially an approximate version of this argument. One finds a prior over all of $\mathbb{R}^n$ whose risk is arbitrarily close to $\sigma^2 c^2/(\sigma^2 + c^2)$ and one then shows that the prior asymptotically concentrates on $\Theta_n(c)$.

Now let us see how minimax theory works for smooth functions.

7.32 Theorem (Pinsker's theorem for Sobolev ellipsoids). *Let*

$$Z_j = \theta_j + \frac{\sigma}{\sqrt{n}}\epsilon_j, \quad j = 1, 2, \ldots \tag{7.33}$$

where $\epsilon_1, \epsilon_2, \ldots \sim N(0,1)$. Assume that $\theta \in \Theta(m,c)$, a Sobolev ellipsoid (recall Definition 7.10). Let R_n denote the minimax risk over $\Theta(m,c)$. Then,

$$\lim_{n \to \infty} n^{2m/(2m+1)} R_n = \left(\frac{\sigma}{\pi}\right)^{2m/(2m+1)} c^{2/(2m+1)} P_m \tag{7.34}$$

where

$$P_m = \left(\frac{m}{m+1}\right)^{2m/(2m+1)} (2m+1)^{1/(2m+1)} \tag{7.35}$$

*is the **Pinsker constant**. Hence, the minimax rate is $n^{-2m/(2m+1)}$, that is,*

$$0 < \lim_{n \to \infty} n^{2m/(2m+1)} R_n < \infty.$$

Here is a more general version of the theorem.

7.36 Theorem (Pinsker's theorem for ellipsoids). *Let*

$$\Theta = \left\{ \theta : \sum_{j=1}^{\infty} a_j \theta_j^2 \le c^2 \right\}.$$

*The set Θ is called an **ellipsoid**. Assume that $a_j \to \infty$ as $j \to \infty$. Let*

$$R_n = \inf_{\widehat{\theta}} \sup_{\theta \in \Theta} R(\widehat{\theta}, \theta)$$

denote the minimax risk and let

$$R_n^L = \inf_{\widehat{\theta} \in \mathcal{L}} \sup_{\theta \in \Theta} R(\widehat{\theta}, \theta)$$

denote the minimax linear risk where $\mathcal{L}$ is the set of linear estimators of the form $\widehat{\theta} = (w_1 Z_1, w_2 Z_2, \ldots)$. Then:

(1) linear estimators are asymptotically minimax: $R_n \sim R_n^L$ as $n \to \infty$;

(2) the minimax linear risk satisfies

$$R_n^L = \frac{\sigma^2}{n} \sum_i \left(1 - \frac{a_i}{\mu}\right)_+$$

where μ solves

$$\frac{\sigma^2}{n} \sum_i a_i(\mu - a_i)_+ = c^2.$$

(3) The linear minimax estimator is $\widehat{\theta}_i = w_i Z_i$ where $w_i = [1 - (a_i/\mu)]_+$.

(4) The linear minimax estimator is Bayes[5] for the prior with independent components such that $\theta_i \sim N(0, \tau_i^2)$, $\tau_i^2 = (\sigma^2/n)(\mu/a_i - 1)_+$.

7.6 Linear Shrinkage and the James–Stein Estimator

Let us now return to model (7.1) and see how we can improve on the MLE using linear estimators. A **linear estimator** is an estimator of the form $\widehat{\theta} = bZ = (bZ_1, \ldots, bZ_n)$ where $0 \le b \le 1$. Linear estimators are **shrinkage estimators** since they **shrink** Z towards the origin. We denote the set of linear shrinkage estimators by $\mathcal{L} = \{bZ : b \in [0,1]\}$.

The risk of a linear estimator is easy to compute. From the basic bias–variance breakdown we have

$$R(bZ, \theta) = (1 - b)^2 ||\theta||_n^2 + nb^2\sigma_n^2 \qquad (7.37)$$

where $||\theta||_n^2 = \sum_{i=1}^n \theta_i^2$. The risk is minimized by taking

$$b_* = \frac{||\theta||_n^2}{n\sigma_n^2 + ||\theta||_n^2}.$$

We call $b_* Z$ the **ideal linear estimator**. The risk of this ideal linear estimator is

$$R(b_* Z, \theta) = \frac{n\sigma_n^2 ||\theta||_n^2}{n\sigma_n^2 + ||\theta||_n^2}. \qquad (7.38)$$

Thus we have proved:

7.39 Theorem.

$$\inf_{\widehat{\theta} \in \mathcal{L}} R(\widehat{\theta}, \theta) = \frac{n\sigma_n^2 ||\theta||_n^2}{n\sigma_n^2 + ||\theta||_n^2}. \qquad (7.40)$$

We can't use the estimator $b_* Z$ because b_* depends on the unknown parameter θ. For this reason we call $R(b_* Z, \theta)$ the **linear oracular risk** since the risk could only be obtained by an "oracle" that knows $||\theta||_n^2$. We shall now show that the **James–Stein estimator** nearly achieves the risk of the ideal oracle.

---

[5] The Bayes estimator minimizes Bayes risk $\int R(\theta, \widehat{\theta})d\pi(\theta)$ for a given prior π.

The James–Stein estimator of θ is defined by

$$\widehat{\theta}^{JS} = \left(1 - \frac{(n-2)\sigma_n^2}{\sum_{i=1}^n Z_i^2}\right) Z. \tag{7.41}$$

We'll see in Theorem 7.48 that this estimator is asymptotically optimal.

7.42 Theorem. *The risk of the James–Stein estimator satisfies the following bound:*

$$R(\widehat{\theta}^{JS}, \theta) \le 2\sigma_n^2 + \frac{(n-2)\sigma_n^2\|\theta\|_n^2}{(n-2)\sigma_n^2 + \|\theta\|_n^2} \le 2\sigma_n^2 + \frac{n\sigma_n^2\|\theta\|_n^2}{n\sigma_n^2 + \|\theta\|_n^2} \tag{7.43}$$

where $\|\theta\|_n^2 = \sum_{i=1}^n \theta_i^2$.

PROOF. Write $\widehat{\theta}^{JS} = Z + g(Z)$ where $g(z) = -(n-2)\sigma_n^2 z / \sum_i z_i^2$. Hence

$$D_i = \frac{\partial g_i}{\partial z_i} = -(n-2)\sigma_n^2 \left(\frac{1}{\sum_i z_i^2} - \frac{2z_i^2}{(\sum_i z_i^2)^2}\right)$$

and

$$\sum_{i=1}^n D_i = -\frac{(n-2)^2\sigma_n^2}{\sum_{i=1}^n z_i^2}.$$

Plugging this into the SURE formula (7.19) yields

$$\widehat{R}(Z) = n\sigma_n^2 - \frac{(n-2)^2\sigma_n^4}{\sum_i Z_i^2}.$$

Hence, the risk is

$$R(\widehat{\theta}^{JS}, \theta) = \mathbb{E}(\widehat{R}(Z)) = n\sigma_n^2 - (n-2)^2\sigma_n^4 \mathbb{E}\left(\frac{1}{\sum_i Z_i^2}\right). \tag{7.44}$$

Now $Z_i^2 = \sigma_n^2((\theta_i/\sigma_n) + \epsilon_i)^2$ and so $\sum_{i=1}^n Z_i^2 \sim \sigma_n^2 W$ where W is noncentral χ^2 with n degrees of freedom and noncentrality parameter $\delta = \sum_{i=1}^n (\theta_i^2/\sigma_n^2)$. Using a well-known result about noncentral χ^2 random variables, we can then write $W \sim \chi_{n+2K}^2$ where $K \sim \text{Poisson}(\delta/2)$. Recall that (for $n > 2$) $\mathbb{E}(1/\chi_n^2) = 1/(n-2)$. So,

$$
\begin{aligned}
\mathbb{E}_\theta\left[\frac{1}{\sum_i Z_i^2}\right] &= \left(\frac{1}{\sigma_n^2}\right)\mathbb{E}\left[\frac{1}{\chi_{n+2K}^2}\right] = \left(\frac{1}{\sigma_n^2}\right)\mathbb{E}\left(E\left[\frac{1}{\chi_{n+2K}^2}\,\Big|\,K\right]\right) \\
&= \left(\frac{1}{\sigma_n^2}\right)\mathbb{E}\left[\frac{1}{n-2+2K}\right] \\
&\ge \left(\frac{1}{\sigma_n^2}\right)\frac{1}{(n-2) + \sigma_n^{-2}\sum_{i=1}^n \theta_i^2} \qquad \text{from Jensen's inequality} \\
&= \frac{1}{(n-2)\sigma_n^2 + \sum_{i=1}^n \theta_i^2}.
\end{aligned}
$$

Substituting into (7.44) we get the first inequality. The second inequality follows from simple algebra. ∎

7.45 Remark. The **modified James–Stein estimator** is defined by

$$\widehat{\theta} = \left(1 - \frac{n\sigma_n^2}{\sum_i Z_i^2}\right)_+ Z \qquad (7.46)$$

where $(a)_+ = \max\{a, 0\}$. The change from $n - 2$ to n leads to a simpler expression and for large n this has negligible effect. Taking the positive part of the shrinkage factor cannot increase the risk. In practice, the modified James–Stein estimator is often preferred.

The next result shows that the James–Stein estimator nearly achieves the risk of the linear oracle.

7.47 Theorem (James–Stein oracle inequality). *Let $\mathcal{L} = \{bZ : b \in \mathbb{R}\}$ denote the class of linear estimators. For all $\theta \in \mathbb{R}^n$,*

$$\inf_{\widehat{\theta} \in \mathcal{L}} R(\widehat{\theta}, \theta) \leq R(\widehat{\theta}^{JS}, \theta) \leq 2\sigma_n^2 + \inf_{\widehat{\theta} \in \mathcal{L}} R(\widehat{\theta}, \theta).$$

PROOF. This follows from (7.38) and Theorem 7.42. ∎

Here is another perspective on the James–Stein estimator. Let $\widehat{\theta} = bZ$. Stein's unbiased risk estimator is $\widehat{R}(Z) = n\sigma_n^2 + 2n\sigma_n^2(b-1) + (b-1)^2 \sum_{i=1}^n Z_i^2$ which is minimized at

$$\widehat{b} = 1 - \frac{n\sigma_n^2}{\sum_{i=1}^n Z_i^2}$$

yielding the estimator

$$\widehat{\theta} = \widehat{b}Z = \left(1 - \frac{n\sigma_n^2}{\sum_{i=1}^n Z_i^2}\right) Z$$

which is essentially the James–Stein estimator.

We can now show that the James–Stein estimator achieves the Pinsker bound (7.29) and so is asymptotically minimax.

7.48 Theorem. *Let $\sigma_n^2 = \sigma^2/n$. The James–Stein estimator is asymptotically minimax, that is,*

$$\lim_{n \to \infty} \sup_{\theta \in \Theta_n(c)} R(\widehat{\theta}^{JS}, \theta) = \frac{\sigma^2 c^2}{\sigma^2 + c^2}.$$

PROOF. Follows from Theorem 7.42 and 7.28. ∎

7.49 Remark. The James–Stein estimator is **adaptive** in the sense that it achieves the minimax bound over $\Theta_n(c)$ without knowledge of the parameter c.

To summarize: the James–Stein estimator is essentially optimal over all linear estimators. Moreover, it is asymptotically optimal over all estimators, not just linear estimators. This also shows that the minimax risk and the linear minimax risk are asymptotically equivalent. This turns out to (sometimes) be a more general phenomenon, as we shall see.

7.7   Adaptive Estimation Over Sobolev Spaces

Theorem 7.32 gives an estimator that is minimax over $\Theta(m, c)$. However, the estimator is unsatisfactory because it requires that we know c and m.

Efromovich and Pinsker (1984) proved the remarkable result that there exists an estimator that is minimax over $\Theta(m, c)$ without requiring knowledge of m or c. The estimator is said to be **adaptively asymptotically minimax**. The idea is to divide the observations into blocks $B_1 = \{Z_1, \ldots, Z_{n_1}\}$, $B_2 = \{Z_{n_1+1}, \ldots, Z_{n_2}\}$, $\ldots$ and then apply a suitable estimation procedure within blocks.

Here is particular block estimation scheme due to Cai et al. (2000). For any real number a let $[a]$ denote the integer part of a. Let $b = 1 + 1/\log n$ and let K_0 be an integer such that $[b^{K_0}] \geq 3$ and $[b^k] - [b^{k-1}] \geq 3$ for $k \geq K_0 + 1$. Let $B_0 = \{Z_i : 1 \leq i \leq [b^{K_0}]\}$ and let $B_k = \{Z_i : [b^{k-1}] < i \leq [b^k]\}$ for $k \geq K_0 + 1$. Let $\widehat{\theta}$ be the estimator obtained by applying the James–Stein estimator within each block B_k. The estimator is taken to be 0 for $i > [b^{K_1}]$ where $K_1 = [\log_b(n)] - 1$.

7.50 Theorem (Cai, Low and Zhao, 2000). *Let $\widehat{\theta}$ be the estimator above. Let $\Theta(m, c) = \{\theta : \sum_{i=1}^{\infty} a_i^2 \theta_i^2 \leq c^2\}$ where $a_1 = 1$ and $a_{2i} = a_{2i+1} = 1 + (2i\pi)^{2m}$. Let $R_n(m, c)$ denote the minimax risk over $\Theta(m, c)$. Then for all $m > 0$ and $c > 0$,*

$$\lim_{n \to \infty} \frac{\sup_{\theta \in \Theta(m,c)} R(\widehat{\theta}, \theta)}{R_n(m, c)} = 1.$$

7.8   Confidence Sets

In this section we discuss the construction of confidence sets for θ^n. It will now be convenient to write θ and Z instead of θ^n and Z^n.

Recall that $\mathcal{B}_n \subset \mathbb{R}^n$ is a $1 - \alpha$ confidence set if

$$\inf_{\theta \in \mathbb{R}^n} \mathbb{P}_\theta(\theta \in \mathcal{B}_n) \geq 1 - \alpha. \tag{7.51}$$

We have written the probability distribution $\mathbb{P}_\theta$ with the subscript θ to emphasize that the distribution depends on θ. Here are some methods for constructing confidence sets.

METHOD I: THE χ^2 CONFIDENCE SET. The simplest confidence set for θ is based on the fact that $||Z - \theta||^2/\sigma_n^2$ has a χ_n^2 distribution. Let

$$\mathcal{B}_n = \left\{ \theta \in \mathbb{R}^n : \ ||Z - \theta||^2 \leq \sigma_n^2\, \chi_{n,\alpha}^2 \right\} \tag{7.52}$$

where $\chi_{n,\alpha}^2$ is the upper α quantile of a χ^2 random variable with n degrees of freedom. It follows immediately that

$$\mathbb{P}_\theta(\theta \in \mathcal{B}_n) = 1 - \alpha, \quad \text{for all } \theta \in \mathbb{R}^n.$$

Hence, (7.51) is satisfied. The expected radius of this ball is $n\sigma_n^2$. We will see that we can improve on this.

IMPROVING THE χ^2 BALL BY PRE-TESTING. Before discussing more complicated methods, here is a simple idea—based on ideas in Lepski (1999)—for improving the χ^2 ball. The methods that follow are generalizations of this method.

Note that the χ^2 ball $\mathcal{B}_n$ has a fixed radius $s_n = \sigma_n \sqrt{n}$. When applied to function estimation, $\sigma_n = O(1/\sqrt{n})$ so that $s_n = O(1)$ and hence the radius of the ball does not even converge to zero as $n \to \infty$. The following construction makes the radius smaller. The idea is to test the hypothesis that $\theta = \theta_0$. If we accept the null hypothesis, we use a smaller ball centered at θ_0. Here are the details.

First, test the hypothesis that $\theta = (0, \ldots, 0)$ using $\sum_i Z_i^2$ as a test statistic. Specifically, reject the null when

$$T_n = \sum_i Z_i^2 > c_n^2$$

and c_n is defined by

$$\mathbb{P}\left(\chi_n^2 > \frac{c_n^2}{\sigma_n^2} \right) = \frac{\alpha}{2}.$$

By construction, the test has type one error rate $\alpha/2$. If Z denotes a N(0,1) random variable then

$$\frac{\alpha}{2} = \mathbb{P}\left(\chi_n^2 > \frac{c_n^2}{\sigma_n^2}\right) = \mathbb{P}\left(\frac{\chi_n^2 - n}{\sqrt{2n}} > \frac{\frac{c_n^2}{\sigma_n^2} - n}{\sqrt{2n}}\right) \approx \mathbb{P}\left(Z > \frac{\frac{c_n^2}{\sigma_n^2} - n}{\sqrt{2n}}\right)$$

implying that

$$c_n^2 \approx \sigma_n^2(n + \sqrt{2n}z_{\alpha/2}).$$

Now we compute the power of this test when $||\theta|| > \Delta_n$ where

$$\Delta_n = \sqrt{2\sqrt{2}z_{\alpha/2}}\, n^{1/4}\sigma_n.$$

Write $Z_i = \theta_i + \sigma_n\epsilon_i$ where $\epsilon_i \sim N(0,1)$. Then,

$$
\begin{aligned}
\mathbb{P}_\theta(T_n > c_n^2) &= \mathbb{P}_\theta\left(\sum_i Z_i^2 > c_n^2\right) = \mathbb{P}_\theta\left(\sum_i (\theta_i + \sigma_n\epsilon_i)^2 > c_n^2\right) \\
&= \mathbb{P}_\theta\left(||\theta||^2 + 2\sigma_n\sum_i \theta_i\epsilon_i + \sigma_n^2\sum_i \epsilon_i^2 > c_n^2\right).
\end{aligned}
$$

Now, $||\theta||^2 + 2\sigma_n\sum_i \theta_i\epsilon_i + \sigma_n^2\sum_i \epsilon_i^2$ has mean $||\theta||^2 + n\sigma_n^2$ and variance $4\sigma_n^2||\theta||^2 + 2n\sigma_n^4$. Hence, with Z denoting a N(0,1) random variable,

$$
\begin{aligned}
\mathbb{P}_\theta(T_n > c_n^2) &\approx \mathbb{P}\left(||\theta||^2 + n\sigma_n^2 + \sqrt{4\sigma_n^2||\theta||^2 + 2n\sigma_n^4}\,Z > c_n^2\right) \\
&\approx \mathbb{P}\left(||\theta||^2 n\sigma_n^2 + \sqrt{4\sigma_n^2||\theta||^2 + 2n\sigma_n^4}\,Z > \sigma_n^2(n + \sqrt{2n}z_{\alpha/2})\right) \\
&= \mathbb{P}\left(Z > \frac{\left(\sqrt{2}z_{\alpha/2} - \frac{||\theta||^2}{\sqrt{n\sigma_n^2}}\right)}{2 + \frac{4||\theta||^2}{n\sigma_n^2}}\right) \geq \mathbb{P}\left(Z > \frac{\left(\sqrt{2}z_{\alpha/2} - \frac{||\theta||^2}{\sqrt{n\sigma_n^2}}\right)}{2}\right) \\
&\geq 1 - \frac{\alpha}{2}
\end{aligned}
$$

since $||\theta|| > \Delta_n$ implies that

$$\frac{\left(\sqrt{2}z_{\alpha/2} - \frac{||\theta||^2}{\sqrt{n\sigma_n^2}}\right)}{2} \geq -z_{\alpha/2}.$$

In summary, the test has type-one error $\alpha/2$ and type-two error no more than $\alpha/2$ for all $||\theta|| > \Delta_n$.

Next we define the confidence procedure as follows. Let $\phi = 0$ if the test accepts and $\phi = 1$ if the test rejects. Define

$$R_n = \begin{cases} \mathcal{B}_n & \text{if } \phi = 1 \\ \left\{\theta : ||\theta|| \leq \Delta_n\right\} & \text{if } \phi = 0. \end{cases}$$

Thus, R_n is a random radius confidence ball. The radius is the same as the χ^2 ball when $\phi = 1$ but when $\phi = 0$, the radius is Δ_n which is much smaller. Let us now verify that the ball has the right coverage.

The noncoverage of this ball when $\theta = (0, \ldots, 0)$ is

$$
\begin{aligned}
\mathbb{P}_0(\theta \notin R) &= \mathbb{P}_0(\theta \notin R, \phi = 0) + \mathbb{P}_0(\theta \notin R, \phi = 1) \\
&\leq 0 + \mathbb{P}_0(\phi = 1) = \frac{\alpha}{2}.
\end{aligned}
$$

The noncoverage of this ball when $\theta \neq (0, \ldots, 0)$ and $||\theta|| \leq \Delta_n$ is

$$
\begin{aligned}
\mathbb{P}_\theta(\theta \notin R) &= \mathbb{P}_\theta(\theta \notin R, \phi = 0) + \mathbb{P}_\theta(\theta \notin R, \phi = 1) \\
&\leq 0 + \mathbb{P}_\theta(\theta \notin B) = \frac{\alpha}{2}.
\end{aligned}
$$

The noncoverage of this ball when $\theta \neq (0, \ldots, 0)$ and $||\theta|| > \Delta_n$ is

$$
\begin{aligned}
\mathbb{P}_\theta(\theta \notin R) &= \mathbb{P}_\theta(\theta \notin R, \phi = 0) + \mathbb{P}_\theta(\theta \notin R, \phi = 1) \\
&\leq \mathbb{P}_\theta(\phi = 0) + \mathbb{P}_\theta(\theta \notin B) \leq \frac{\alpha}{2} + \frac{\alpha}{2} = \alpha.
\end{aligned}
$$

In summary, by testing whether θ is close to $(0, \ldots, 0)$ and using a smaller ball centered at $(0, \ldots, 0)$ when the test accepts, we get a ball with proper coverage and whose radius is sometimes smaller than the χ^2 ball. The message is that:

> a random radius confidence ball can have an expected radius that is smaller than a fixed confidence ball at some points in the parameter space.

The next section generalizes this idea.

METHOD II: THE BARAUD CONFIDENCE SET. Here we discuss the method due to Baraud (2004) which builds on Lepski (1999), as discussed above. We begin with a class $\mathcal{S}$ of linear subspaces of $\mathbb{R}^n$. Let Π_S denote the projector onto S. Thus, for any vector $Z \in \mathbb{R}^n$, $\Pi_S Z$ is the vector in S closest to Z.

For each subspace S, we construct a ball $\mathcal{B}_S$ of radius ρ_S centered at an estimator in S, namely,

$$
\mathcal{B}_S = \left\{ \theta : ||\theta - \Pi_S Z|| \leq \rho_S \right\}. \tag{7.53}
$$

For each $S \in \mathcal{S}$, we test whether θ is close to S using $||Z - \Pi_S Z||$ as a test statistic. We then take the smallest confidence ball $\mathcal{B}_S$ among all unrejected subspaces S. The key to making this work is this: the radius ρ_S is chosen so that

$$\max_\theta \mathbb{P}_\theta(S \text{ is not rejected and } \theta \notin \mathcal{B}_S) \leq \alpha_S \qquad (7.54)$$

where $\sum_{S \in \mathcal{S}} \alpha_S \leq \alpha$. The resulting confidence ball has coverage at least $1 - \alpha$ since

$$
\begin{aligned}
\max_\theta \mathbb{P}_\theta(\theta \notin \mathcal{B}) \;&\leq\; \sum_S \max_\theta \mathbb{P}_\theta(S \text{ is not rejected and } \theta \notin \mathcal{B}_S) \\
&=\; \sum_S \alpha_S \leq \alpha.
\end{aligned}
$$

We will see that the n-dimensional maximization over $\theta \in \mathbb{R}^n$ can be reduced to a one-dimensional maximization since the probabilities only depend on θ through the quantity $z = ||\theta - \Pi_S \theta||$.

The confidence set has coverage $1 - \alpha$ even if θ is not close to one of the subspaces in $\mathcal{S}$. However, if it is close to one of the subspaces in $\mathcal{S}$, then the confidence ball will be smaller than the χ^2 ball.

For example, suppose we expand a function $f(x) = \sum_j \theta_j \phi_j(x)$ in a basis, as in Section 7.3. Then, the θ_is correspond to the coefficients of f in this basis. If the function is smooth, then we expect that θ_i will be small for large i. Hence, θ might be well approximated by a vector of the form $(\theta_1, \ldots, \theta_m, 0, \ldots, 0)$. This suggests that we could test whether θ is close to the subspace S_m of the vectors of the form $(\theta_1, \ldots, \theta_m, 0, \ldots, 0)$, for $m = 0, \ldots, n$. In this case we would take the class of subspaces to be $\mathcal{S} = \{S_0, \ldots, S_n\}$.

Before we proceed with the details, we need some notation. If $X_j \sim N(\mu_j, 1)$, $j = 1, \ldots, k$ are IID, then $T = \sum_{j=1}^k X_j^2$ has a noncentral χ^2 distribution with noncentrality parameter $d = \sum_j \mu_j^2$ and k degrees of freedom and we write $T \sim \chi^2_{d,k}$. Let $G_{d,k}$ denote the CDF of this random variable and let $q_{d,k}(\alpha) = G_{d,k}^{-1}(1 - \alpha)$ denote the upper α quantile. By convention, we define $q_{d,k}(\alpha) = -\infty$ for $\alpha \geq 1$.

Let $\mathcal{S}$ be a finite collection of linear subspaces of $\mathbb{R}^n$. We assume that $\mathbb{R}^n \in \mathcal{S}$. Let $d(S)$ be the dimension of $S \in \mathcal{S}$ and let $e(S) = n - d(S)$. Fix $\alpha \in (0,1)$ and $\gamma \in (0,1)$ where $\gamma < 1 - \alpha$. Let

$$\mathcal{A} = \left\{ S : \frac{||Z - \Pi_S Z||^2}{\sigma_n^2} \leq c(S) \right\} \qquad (7.55)$$

where

$$c(S) = q_{0,e(S)}(\gamma). \qquad (7.56)$$

Think of $||Z - \Pi_S Z||^2$ as a test statistic for the hypothesis that $\theta \in S$. Then $\mathcal{A}$ is the set of nonrejected subspaces. Note that $\mathcal{A}$ always includes the subspace $S = \mathbb{R}^n$ since, when $S = \mathbb{R}^n$, $\Pi_S Z = Z$ and $||Z - \Pi_S Z||^2 = 0$.

Let $(\alpha_S : S \in \mathcal{S})$ be a set of numbers such that $\sum_{S \in \mathcal{S}} \alpha_S \leq \alpha$. Now define the ρ_S as follows:

$$\rho_S^2 = \sigma_n^2 \times \begin{cases} \inf_{z>0}\left\{ G_{z,n}(q_{0,n}(\gamma)) \leq \alpha_S \right\} & \text{if } d(S) = 0 \\[2ex] \sup_{z>0}\left\{ z + q_{0,d(S)}\left(\frac{\alpha_S}{G_{z,e(S)}(c(S))} \right) \right\} & \text{if } 0 < d(S) < n \\[2ex] \rho_S^2 = \sigma_n^2 q_{0,n}(\alpha_S) & \text{if } d(S) = n. \end{cases} \tag{7.57}$$

Define

$$\widehat{S} = \operatorname{argmin}_{S \in \mathcal{A}} \rho_S,$$

$\widehat{\theta} = \Pi_{\widehat{S}} Z$, and $\widehat{\rho} = \rho_{\widehat{S}}$. Finally, define

$$\mathcal{B}_n = \left\{ \theta \in \mathbb{R}^n : ||\theta - \widehat{\theta}||^2 \leq \widehat{\rho}^2 \right\}. \tag{7.58}$$

7.59 Theorem (Baraud 2004). *The set $\mathcal{B}_n$ defined in (7.58) is a valid confidence set:*

$$\inf_{\theta \in \mathbb{R}^n} \mathbb{P}_\theta(\theta \in \mathcal{B}_n) \geq 1 - \alpha. \tag{7.60}$$

PROOF. Let $\mathcal{B}_S = \{\theta : ||\theta - \Pi_S Z||^2 \leq \rho_S^2\}$. Then,

$$\begin{aligned} \mathbb{P}_\theta(\theta \notin \mathcal{B}_n) &\leq \mathbb{P}_\theta(\theta \notin \mathcal{B}_S \text{ for some } S \in \mathcal{A}) \\ &\leq \sum_S \mathbb{P}_\theta(||\theta - \Pi_S Z|| > \rho_S, \ \widehat{S} \in \mathcal{A}) \\ &= \sum_S \mathbb{P}_\theta\left(||\theta - \Pi_S Z|| > \rho_S, \ ||Z - \Pi_S Z||^2 \leq c(S)\sigma_n^2\right). \end{aligned}$$

Since $\sum_S \alpha_S \leq \alpha$, it suffices to show that $a(S) \leq \alpha_S$ for all $S \in \mathcal{S}$, where

$$a(S) \equiv \mathbb{P}_\theta\left(||\theta - \Pi_S Z|| > \rho_S, \ ||Z - \Pi_S Z||^2 \leq \sigma_n^2 c(S) \right). \tag{7.61}$$

When $d(S) = 0$, $\Pi_S Z = (0, \dots, 0)$. If $||\theta|| \leq \rho_S$ then $a(0) = 0$ which is less than α_S. If $||\theta|| > \rho_S$, then

$$\begin{aligned} a(S) &= \mathbb{P}_\theta\left(\sum_{i=1}^n Z_i^2 \leq \sigma_n^2 q_{0,n}(\gamma) \right) \\ &= G_{||\theta||^2/\sigma_n^2, n}(q_{0,n}(\gamma)) \leq G_{\rho_0^2/\sigma_n^2, n}(q_{0,n}(\gamma)) \\ &\leq \alpha_S \end{aligned}$$

since $G_{d,n}(u)$ is decreasing in z for all u and from the definition of ρ_0^2.

Now consider the case where $0 < d(S) < n$. Let

$$A = \frac{\|\theta - \Pi_S Z\|^2}{\sigma_n^2} = z + \sum_{j=1}^{m} \epsilon_j^2, \qquad B = \frac{\|\hat{\theta} - Z\|^2}{\sigma_n^2}$$

where $z = \|\theta - \Pi_S \theta\|^2 / \sigma_n^2$. Then A and B are independent, $A \sim z + \chi^2_{0,d(S)}$, and $B \sim \chi^2_{z,e(S)}$. Hence,

$$
\begin{aligned}
a(S) &= \mathbb{P}_\theta \left(A > \frac{\rho_m^2}{\sigma_n^2}, \ B < c(S) \right) \\
&= \mathbb{P}_\theta \left(z + \chi^2_{d(S)} > \frac{\rho_S^2}{\sigma_n^2}, \ \chi^2_{z,e(S)} < c(S) \right) \qquad (7.62) \\
&= \left(1 - G_{0,d(S)} \left(\frac{\rho_S^2}{\sigma_n^2} - z \right) \right) \times G_{z,e(S)} \big(c(S) \big). \qquad (7.63)
\end{aligned}
$$

From the definition of ρ_S^2,

$$\frac{\rho_S^2}{\sigma_n^2} - z \geq q_{0,d(S)} \left(\frac{\alpha_S}{G_{z,e(S)}(c(S))} \wedge 1 \right)$$

and hence,

$$
\begin{aligned}
1 - G_{0,d(S)} \left(\frac{\rho_S^2}{\sigma_n^2} - z \right) &\leq 1 - G_{0,d(S)} \left(q_{0,d(S)} \left(\frac{\alpha_S}{G_{z,e(S)}(c(S))} \right) \right) \\
&= \frac{\alpha_S}{G_{z,e(S)}(c(S))}. \qquad (7.64)
\end{aligned}
$$

It then follows (7.63) and (7.64) that $a(S) \leq \alpha_S$.

For the case $d(S) = n$, $\Pi_S Z = Z$, and $\|\theta - \Pi_S Z\|^2 = \sigma_n^2 \sum_{i=1}^{n} \epsilon_i^2 \overset{d}{=} \sigma_n^2 \chi_n^2$ and so

$$a(S) = \mathbb{P}_\theta \big(\sigma_n^2 \chi_n^2 > q_{0,n}(\alpha_S) \sigma_n^2 \big) = \alpha_S$$

by the definition of $q_{0,n}$. ∎

When σ_n is unknown we estimate the variance using one of the methods discussed in Chapter 5 and generally the coverage is only asymptotically correct. To see the effect of having uncertainty about σ_n, consider the idealized case where σ_n is known to lie with certainty in the interval $I = [\sqrt{1 - \eta_n} \tau_n, \tau_n]$. (In practice, we would construct a confidence interval for σ and adjust the level α of the confidence ball appropriately.) In this case, the radii ρ_S are now defined by:

$$\rho_S^2 = \begin{cases} \inf_{z>0}\left\{\sup_{\sigma_n \in I} G_{z/\sigma_n^2, n}(q_{0,n}(\gamma)\tau_n^2/\sigma_n^2) \leq \alpha_S\right\} & \text{if } d(S) = 0 \\ \sup_{\tau > 0, \sigma_u \leq I}\left\{z\sigma_n^2 + \sigma_n^2 q_{0,d(S)}(h_S(z, \sigma_n))\right\} & \text{if } 0 < d(S) < n \\ q_{0,n}(\alpha_S)\tau_n^2 & \text{if } d(S) = n \end{cases}$$

(7.65)

where

$$h_S(z, \sigma) = \frac{\alpha_S}{G_{z,e(S)}\left(G_{z,e(S)}(q_{0,e(S)}(\gamma)\tau_n^2/\sigma^2)\right)} \qquad (7.66)$$

and $\mathcal{A}$ is now defined by

$$\mathcal{A} = \left\{S \in \mathcal{S} : \ ||Z - \Pi_S Z||^2 \leq q_{0,e(S)}(\gamma)\tau_n^2\right\}. \qquad (7.67)$$

BERAN–DÜMBGEN–STEIN PIVOTAL METHOD. Now we discuss a different approach due to Stein (1981) and developed further by Li (1989), Beran and Dümbgen (1998), and Genovese and Wasserman (2005). The method is simpler than the Baraud–Lepski approach but it uses asymptotic approximations. This method is considered in more detail in the next chapter but here is the basic idea.

Consider nested subsets $\mathcal{S} = \{S_0, S_1, \ldots, S_n\}$ where

$$S_j = \left\{\theta = (\theta_1, \ldots, \theta_j, 0, \ldots, 0) : \ (\theta_1, \ldots, \theta_j) \in \mathbb{R}^j\right\}.$$

Let $\widehat{\theta}_m = (Z_1, \ldots, Z_m, 0, \ldots, 0)$ denote the estimator under model S_m. The loss function is

$$L_m = ||\widehat{\theta}_m - \theta||^2.$$

Define the **pivot**

$$V_m = \sqrt{n}(L_m - \widehat{R}_m) \qquad (7.68)$$

where $\widehat{R}_m = m\sigma_n^2 + \sum_{j=m+1}^{n}(Z_j^2 - \sigma_n^2)$ is SURE. Let $\widehat{m}$ minimize $\widehat{R}_m$ over m. Beran and Dümbgen (1998) show that $V_{\widehat{m}}/\widehat{\tau} \rightsquigarrow N(0, 1)$ where

$$\tau_m^2 = \mathbb{V}(V_m) = 2n\sigma_n^2\left(n\sigma_n^2 + 2\sum_{j=m+1}^{n}\theta_j^2\right)$$

and

$$\widehat{\tau}^2 = 2n\sigma_n^2\left(n\sigma_n^2 + 2\sum_{j=\widehat{m}+1}^{n}(Z_j^2 - \sigma_n^2)\right).$$

Let

$$r_n^2 = \widehat{R}_m + \frac{\widehat{\tau} z_\alpha}{\sqrt{n}}$$

and define

$$\mathcal{B}_n = \left\{ \theta \in \mathbb{R}^n : \ ||\theta_m - \widehat{\theta}||^2 \le r_n^2 \right\}.$$

Then,

$$
\begin{aligned}
\mathbb{P}_\theta(\theta \in \mathcal{B}_n) \ &= \ \mathbb{P}_\theta(||\theta - \widehat{\theta}||^2 \le r_n^2) = \mathbb{P}_\theta(L_m \le r_n^2) \\
&= \ \mathbb{P}_\theta\left(L_m \le \widehat{R}_m + \frac{\widehat{\tau} z_\alpha}{\sqrt{n}} \right) = \mathbb{P}_\theta\left(\frac{V_{\widehat{m}}}{\widehat{\tau}} \le z_\alpha \right) \\
&\to \ 1 - \alpha.
\end{aligned}
$$

A practical problem with this method is that r_n^2 can be negative. This is due to the presence of the term $\sum_{j=m+1}^{n}(Z_j^2 - \sigma_n^2)$ in $\widehat{R}$ and τ. We deal with this by replacing such terms with $\max\{\sum_{j=m+1}^{n}(Z_j^2 - \sigma_n^2), 0\}$. This can lead to over-coverage but at least leads to well-defined radii.

7.69 Example. Consider nested subsets $\mathcal{S} = \{S_0, S_1, \ldots, S_n\}$ where $S_0 = \{(0, \ldots, 0)\}$ and

$$S_j = \left\{ \theta = (\theta_1, \ldots, \theta_j, 0, \ldots, 0) : \ (\theta_1, \ldots, \theta_j) \in \mathbb{R}^j \right\}.$$

We take $\alpha = 0.05$, $n = 100$, $\sigma_n = 1/\sqrt{n}$, and $\alpha_S = \alpha/(n+1)$ for all S so that $\sum \alpha_S = \alpha$ as required. Figure 7.2 shows ρ_S versus the dimension of S for $\gamma = 0.05, 0.15, 0.50, 0.90$. The dotted line is the radius of the χ^2 ball. One can show that

$$\frac{\rho_0}{\rho_n} = O\left(n^{-1/4} \right) \tag{7.70}$$

which shows that shrinking towards lower-dimensional models leads to smaller confidence sets. There is an interesting tradeoff. Setting γ large makes ρ_0 small leading to a potentially smaller confidence ball. However, making γ large increases the set $\mathcal{A}$ which diminishes the chances of choosing a small ρ. We simulated under the model $\theta = (10, 10, 10, 10, 10, 0, \ldots, 0)$. See Table 7.1 for a summary. In this example, the pivotal method seems to perform the best. ∎

7.9   Optimality of Confidence Sets

How small can we make the confidence set while still maintaining correct coverage? In this section we will see that if $\mathcal{B}_n$ is a confidence ball with radius

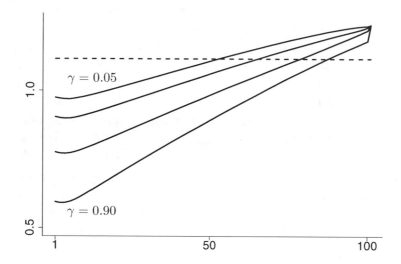

FIGURE 7.2. Constants ρ_S from Example 7.69. The horizontal axis is the dimension of the submodel. The four curves show ρ_S for $\gamma = 0.05, 0.15, 0.50, 0.90$. The highest curve corresponds to $\gamma = 0.05$ and the curves get lower as γ increases. The dotted line is the radius of the χ^2 ball.

Method	Coverage	Radius
χ^2	0.950	1.115
Baraud ($\gamma = 0.90$)	1.000	0.973
($\gamma = 0.50$)	1.000	0.904
($\gamma = 0.15$)	1.000	0.779
($\gamma = 0.05$)	0.996	0.605
Pivotal	0.998	0.582

TABLE 7.1. Simulation results from Example 7.69 based on 1000 simulations.

s_n then $E_\theta(s_n) \geq C_1 \sigma_n n^{1/4}$ for **every** θ and $E_\theta(s_n) \geq C_2 \sigma_n n^{1/2}$ for **some** θ. Here, C_1 and C_2 are positive constants. The χ^2 ball has radius $\sigma_n n^{1/2}$ for all θ. This suggests that the χ^2 ball can be improved upon, and indeed, the Baraud confidence ball can achieve the faster $\sigma_n n^{1/4}$ rate at some points in the parameter space. We will provide some of the details in this section. But first, let us compare this with point estimation.

From Theorem 7.32, the optimal rate of convergence of a point estimator over a Sobolev space of order m is $n^{-2m/(2m+1)}$. According to Theorem 7.50, we can construct estimators that achieve this rate, without prior knowledge of m. This raises the following questions: Can we construct confidence balls that adaptively achieve this optimal rate? The short answer is no. Robins and van der Vaart (2005), Juditsky and Lambert-Lacroix (2003), and Cai and Low (2005) show that some degree of adaptivity is possible for confidence sets but the amount of adaptivity is quite restricted. Without any smoothness assumptions, we see from our comments above that the fastest rate of convergence one can attain is $\sigma_n n^{1/4}$ which is of order $O(n^{-1/4})$ when $\sigma_n = \sigma/\sqrt{n}$.

Turning to the details, we begin with the following Theorem due to Li (1989).

7.71 Theorem (Li 1989). *Let $\mathcal{B}_n = \{\theta^n \in \mathbb{R}^n : ||\widehat{\theta}^n - \theta^n|| \leq s_n\}$ where $\widehat{\theta}^n$ is any estimator of θ^n and $s_n = s_n(Z^n)$ is the radius of the ball. Suppose that*

$$\liminf_{n \to \infty} \inf_{\theta^n \in \mathbb{R}^n} \mathbb{P}_{\theta^n}(\theta^n \in \mathcal{B}_n) \geq 1 - \alpha. \tag{7.72}$$

Then for any sequence θ^n and any $c_n \to 0$,

$$\limsup_{n \to \infty} \mathbb{P}_{\theta^n}(s_n \leq c_n \sigma_n n^{1/4}) \leq \alpha. \tag{7.73}$$

Finite sample results are available from Baraud (2004) and Cai and Low (2005). For example, we have the following result, whose proof is in the appendix.

7.74 Theorem (Cai and Low 2004). *Assume the model (7.1). Fix $0 < \alpha < 1/2$. Let $\mathcal{B}_n = \{\theta : ||\widehat{\theta} - \theta|| \leq s_n\}$ be such that*

$$\inf_{\theta \in \mathbb{R}^n} \mathbb{P}_\theta(\theta \in \mathcal{B}_n) \geq 1 - \alpha.$$

Then, for every $0 < \epsilon < (1/2) - \alpha$,

$$\inf_{\theta \in \mathbb{R}^n} \mathbb{E}_\theta(s_n) \geq \sigma_n(1 - 2\alpha - 2\epsilon)n^{1/4}(\log(1 + \epsilon^2))^{1/4}. \tag{7.75}$$

In particular, if $\sigma_n = \sigma/\sqrt{n}$, then

$$\inf_{\theta \in \mathbb{R}^n} \mathbb{E}_\theta(s_n) \geq \frac{C}{n^{1/4}} \tag{7.76}$$

where $C = \sigma(1 - 2\alpha - 2\epsilon)(\log(1 + \epsilon^2))^{1/4}$.

The lower bound in the above theorem cannot be attained everywhere, as the next result shows.

7.77 Theorem (Cai and Low 2004). *Assume the model (7.1). Fix $0 < \alpha < 1/2$. Let $\mathcal{B}_n = \{\theta : ||\widehat{\theta} - \theta|| \leq s_n\}$ be such that*

$$\inf_{\theta \in \mathbb{R}^n} \mathbb{P}_\theta(\theta \in \mathcal{B}_n) \geq 1 - \alpha.$$

Then, for every $0 < \epsilon < (1/2) - \alpha$,

$$\sup_{\theta \in \mathbb{R}^n} \mathbb{E}_\theta(s_n) \geq \epsilon \sigma_n z_{\alpha + 2\epsilon} \sqrt{n} \sqrt{\frac{\epsilon}{1 - \alpha - \epsilon}}. \tag{7.78}$$

In particular, if $\sigma_n = \sigma/\sqrt{n}$, then

$$\sup_{\theta \in \mathbb{R}^n} \mathbb{E}_\theta(s_n) \geq C \tag{7.79}$$

where $C = \epsilon z_{\alpha + 2\epsilon} \sqrt{\epsilon/(1 - \alpha - \epsilon)}$.

Despite these pessimistic sounding results, there is some potential for adaptation since the infimum in Theorem 7.74 is smaller than the supremum in Theorem 7.77. For example, the χ^2 ball has radius $O(\sigma_n \sqrt{n})$ but the lower bound in the above theorem is $O(\sigma_n n^{1/4})$ suggesting that we can do better than the χ^2 ball. This was the motivation for the Baraud and pivotal confidence sets. The Baraud confidence set does have a certain type of adaptivity: if $\theta \in S$ then $\widehat{\rho} \leq \rho_S$ with high probability. This follows easily from the way that the ball is defined. Let us formalize this as a lemma.

7.80 Lemma. *Define $\mathcal{S}$, α, γ and $(\rho_S : S \in \mathcal{S})$ as in Theorem 7.59. For each $S \in \mathcal{S}$,*

$$\inf_{\theta \in S} \mathbb{P}_\theta(\widehat{\rho} \leq \rho_S) \geq 1 - \gamma. \tag{7.81}$$

Baraud also gives the following results which show that his construction is essentially optimal. The first result gives a lower bound on any adaptive confidence ball. The result after that shows that the radius ρ_S of his confidence set essentially achieves this lower bound.

7.82 Theorem (Baraud 2004). *Suppose that $\widehat{\theta} = \widehat{\theta}(Z)$ and $r = r(Z)$ are such that $\mathcal{B} = \{\theta : \|\theta - \widehat{\theta}\|^2 \leq r^2\}$ is a $1 - \alpha$ confidence ball. Also suppose that $2\alpha + \gamma < 1 - e^{-1/36}$ and that $d(S) \leq n/2$. If*

$$\inf_{\theta \in S} \mathbb{P}_\theta(r \leq r_S) \geq 1 - \gamma \tag{7.83}$$

then, for some $C = C(\alpha, \gamma) > 0$,

$$r_S^2 \geq C\sigma_n^2 \max\{d(S), \sqrt{n}\}. \tag{7.84}$$

Taking S to consist of a single point yields the same result as Theorem 7.74 and taking $S = \mathbb{R}^n$ yields the same result as Theorem 7.77.

7.85 Theorem (Baraud 2004). *Define $\mathcal{S}$, α, γ and $(\rho_S : S \in \mathcal{S})$ as in Theorem 7.59. Assume that $d(S) \leq n/2$ for every $S \in \mathcal{S}$ except for $S = \mathbb{R}^n$. There exists a universal constant $C > 0$ such that*

$$\rho_S^2 \leq C\,\sigma_n^2 \max\{d(S), \sqrt{n\log(1/\alpha_S)}, \log(1/\alpha_S)\}. \tag{7.86}$$

When σ_n is only known to lie in an interval $I = [\sqrt{1 - \eta_n}\tau_n, \tau_n]$, Baraud shows that the lower bound (7.84) becomes

$$r_S^2 \geq C\tau_n^2 \max\{\eta_n n/2, d(S)(1 - \eta_n), \sqrt{n - d(S)}(1 - \eta_n)\} \tag{7.87}$$

which shows that information about σ is crucial. Indeed, the best we realistically could hope for is to know σ^2 up to order $\eta_n = O(n^{-1/2})$ in which case the lower bound is of order $\max\{\sqrt{n}, d(S)\}$.

7.10   Random Radius Bands?

We have seen that random radius confidence balls can be adaptive in the sense that they can be smaller than fixed radius confidence balls at some points in the parameter space. Is the same true for confidence bands? The answer is no, as follows from results in Low (1997). Actually, Low considers estimating a density f at a single point x but essentially the same results apply to regression and to confidence bands. He shows that any random radius confidence interval for $f(x)$ must have expected width at least as large as a fixed width confidence interval. Thus, there is a qualitative difference between constructing a confidence ball versus a confidence band.

Similar comments apply for other norms. The L_p norm is defined by

$$\|\theta\|_p = \begin{cases} (\sum_i |\theta_i|^p)^{1/p} & p < \infty \\ \max_i |\theta_i| & p = \infty. \end{cases}$$

Confidence bands can be thought of as L_∞ confidence balls. It can be shown that confidence balls in the L_p norm with $2 < p < \infty$ fall in between the two extremes of L_2 and L_∞ in the sense that they have some adaptivity, but not as much as in the L_2 norm. Similar comments apply to hypothesis testing; see Ingster and Suslina (2003).

7.11 Penalization, Oracles and Sparsity

Consider again the many Normal means problem

$$Z_i \sim \theta_i + \sigma_n \epsilon_i, \quad i = 1, \ldots, n.$$

If we choose $\widehat{\theta}$ to minimize the sums of squares $\sum_{i=1}^{n}(Z_i - \widehat{\theta}_i)^2$, we get the MLE $\widehat{\theta} = Z = (Z_1, \ldots, Z_n)$. If instead we minimize a penalized sums of squares, we get different estimators.

7.88 Theorem. *Let $J : \mathbb{R}^n \to [0, \infty)$, $\lambda \geq 0$ and define the **penalized sums of squares***

$$M = \sum_{i=1}^{n}(Z_i - \theta_i)^2 + \lambda \, J(\theta).$$

Let $\widehat{\theta}$ minimize M. If $\lambda = 0$ then $\widehat{\theta} = Z$. If $J(\theta) = \sum_{i=1}^{n} \theta_i^2$ then $\widehat{\theta}_i = Z_i/(1+\lambda)$ which is a linear shrinkage estimator. If $J(\theta) = \sum_{i=1}^{n} |\theta_i|$ then $\widehat{\theta}$ is the soft-thresholding estimator (7.21). If $J(\theta) = \#\{\theta_i : \theta_i \neq 0\}$ then $\widehat{\theta}$ is the hard-thresholding estimator (7.23).

Thus we see that linear shrinkage, soft thresholding and hard thresholding are all special cases of one general approach. The case of the L_1 penalty $\sum_{i=1}^{n} |\theta_i|$ is especially interesting. According to Theorem 7.88, the estimator that minimizes

$$\sum_{i=1}^{n}(Z_i - \widehat{\theta}_i)^2 + \lambda \sum_{i=1}^{n} |\theta_i| \tag{7.89}$$

is the soft-threshold estimator $\widehat{\theta}_\lambda = (\widehat{\theta}_{\lambda,1}, \ldots, \widehat{\theta}_{\lambda,n})$ where

$$\widehat{\theta}_{i,\lambda} = \operatorname{sign}(Z_i)(|Z_i| - \lambda)_+.$$

The criterion (7.89) arises in variable selection for linear regression under the name lasso (Tibshirani (1996)) and in signal processing under the name basis pursuit (Chen et al. (1998)). We will see in Chapter 9 that soft thresholding also plays an important role in wavelet methods.

To get more insight on soft thresholding, we consider a result from Donoho and Johnstone (1994). Consider estimating θ_i and suppose we use either Z_i or 0 as an estimator. Such an estimator might be appropriate if we think the vector θ is sparse in the sense that it has many zeroes. The risk of Z_i is σ_n^2 and the risk of 0 is θ_i^2. Imagine an **oracle** that knows when Z_i has better risk and when 0 has better risk. The risk of the oracles estimator is $\min\{\sigma_n^2, \theta_i^2\}$. The risk for estimating the whole vector θ is

$$R_{\text{oracle}} = \sum_{i=1}^{n} \min\{\sigma_n^2, \theta_i^2\}.$$

Donoho and Johnstone (1994) showed that soft thresholding gives an estimator that comes close to the oracle.

7.90 Theorem (Donoho and Johnstone 1994). *Let* $\lambda = \sigma_n\sqrt{2\log n}$. *Then, for every* $\theta \in \mathbb{R}^n$,

$$\mathbb{E}_\theta||\widehat{\theta}_\lambda - \theta||^2 \le (2\log n + 1)(\sigma_n^2 + R_{\text{oracle}}).$$

Moreover, no estimator can get substantially closer to the oracle in the sense that, as $n \to \infty$,

$$\inf_{\widehat{\theta}} \sup_{\theta \in \mathbb{R}^n} \frac{\mathbb{E}_\theta||\widehat{\theta} - \theta||^2}{\sigma_n^2 + R_{\text{oracle}}} \sim 2\log n. \tag{7.91}$$

Consider now a sparse vector θ that is 0 except for k large components, where $k << n$. Then, $R_{\text{oracle}} = k\sigma_n^2$. In function estimation problems, we will see in the next chapter that $\sigma_n^2 = O(1/n)$ and hence $R_{\text{oracle}} = O(k/n)$ which is small in sparse cases (k small).

7.12   Bibliographic Remarks

The idea of reducing nonparametric models to Normal means models (or the white noise model in the appendix) dates back at least to Ibragimov and Has'minskii (1977), Efromovich and Pinsker (1982), and others. See Brown and Low (1996), Nussbaum (1996a) for examples of recent results in this area. A thorough treatment of Normal decision theory and its relation to nonparametric problems is contained in Johnstone (2003). There is also a substantial literature on hypothesis testing in this framework. Many of the results are due to Ingster and are summarized in Ingster and Suslina (2003).

7.13   Appendix

The White Noise Model. Regression is also connected with the **white noise model**. Here is a brief description. Recall that a standard Brownian motion $W(t)$, $0 \le t \le 1$ is a random function such that $W(0) = 0$, $W(s+t) - W(s) \sim N(0, t)$ and, $W(v) - W(u)$ is independent of $W(t) - W(s)$ for $0 \le u \le v \le s \le t$. You can think of W as a continuous version of a random walk. Let $Z_i = f(i/n) + \sigma \epsilon_i$ with $\epsilon_i \sim N(0, 1)$. For $0 \le t \le 1$, define

$$Z_n(t) = \frac{1}{n} \sum_{i=1}^{[nt]} Z_i = \frac{1}{n} \sum_{i=1}^{[nt]} f(i/n) + \frac{\sigma}{\sqrt{n}} \frac{1}{\sqrt{n}} \sum_{i=1}^{[nt]} Z_i.$$

The term $\frac{1}{n} \sum_{i=1}^{[nt]} f(i/n)$ converges to $\int_0^t f(s)ds$ as $n \to \infty$. The term $n^{-1/2} \sum_{i=1}^{[nt]} Z_i$ converges to a standard Brownian motion. (For any fixed t, this is just an application of the central limit theorem.) Thus, asymptotically we can write

$$Z(t) = \int_0^t f(s)ds + \frac{\sigma}{\sqrt{n}} W(t).$$

This is called the **standard white noise model**, often written in differential form as

$$dZ(t) = f(t)dt + \frac{\sigma}{\sqrt{n}} dW(t) \tag{7.92}$$

where $dW(t)$ is the white noise process.[6]

Let $\phi_1, \phi_2, \ldots$ be an orthonormal basis for $L_2(0, 1)$ and write $f(x) = \sum_{i=1}^{\infty} \theta_i \phi_i(x)$ where $\theta_i = \int f(x)\phi_i(x)dx$. Multiply (7.92) by ϕ_j and integrate. This yields $Z_i = \theta_i + (\sigma/\sqrt{n})\epsilon_i$ where $Z_i = \int \phi_i(t)dZ(t)$ and $\epsilon_i = \int \phi_i(t)dW(t) \sim N(0, 1)$. We are back to the Normal means problem. A more complicated argument can be used to relate density estimation to the white noise model as in Nussbaum (1996a).

Weak Differentiability. Let f be integrable on every bounded interval. Then f is **weakly differentiable** if there exists a function f' that is integrable on every bounded interval, such that

$$\int_x^y f'(s)ds = f(y) - f(x)$$

whenever $x \le y$. We call f' the weak derivative of f. An equivalent condition is that for every ϕ that is compactly supported and infinitely differentiable,

$$\int f(s)\phi'(s)ds = -\int f'(s)\phi(s)ds.$$

---

[6]Intuitively, think of $dW(t)$ as a vector of Normals on a very fine grid.

See Härdle et al. (1998), page 72.

Proof of Pinsker's Theorem (Theorem 7.28). (Following Nussbaum (1996b).)
We will need to use Bayes estimators, which we now review. Let π_n be a prior
for θ^n. The **integrated risk** is defined to be $B(\widehat{\theta}, \pi_n) = \int R(\widehat{\theta}^n, \theta^n) d\pi_n(\theta^n) = \mathbb{E}_{\pi_n} \mathbb{E}_\theta L(\widehat{\theta}, \theta)$. The **Bayes estimator** $\widehat{\theta}_{\pi_n}$ minimizes the Bayes risk:

$$B(\pi_n) = \inf_{\widehat{\theta}} B(\widehat{\theta}^n, \pi_n). \tag{7.93}$$

An explicit formula for the Bayes estimator is

$$\widehat{\theta}_{\pi_n}(y) = \mathrm{argmin}_a \mathbb{E}\big(L(a, \theta) \mid Z^n\big).$$

In the case of squared error loss $L(a, \theta) = ||a - \theta||_n^2$, the Bayes estimator is
$\widehat{\theta}_{\pi_n}(y) = \mathbb{E}(\theta | Z^n)$.

Let $\Theta_n = \Theta_n(c)$. Let

$$R_n = \inf_{\widehat{\theta}} \sup_{\theta \in \Theta_n} R(\widehat{\theta}, \theta)$$

denote the minimax risk. We will find an upper bound and a lower bound on
the risk.

UPPER BOUND. Let $\widehat{\theta}_j = c^2 Z_j / (\sigma^2 + c^2)$. The bias of this estimator is

$$\mathbb{E}_\theta(\widehat{\theta}_j) - \theta_j = -\frac{\sigma^2 \theta_j}{\sigma^2 + c^2}$$

and the variance is

$$\mathbb{V}_\theta(\widehat{\theta}_j) = \left(\frac{c^2}{c^2 + \sigma^2}\right)^2 \sigma_n^2 = \left(\frac{c^2}{c^2 + \sigma^2}\right)^2 \frac{\sigma^2}{n}$$

and hence the risk is

$$
\begin{aligned}
\mathbb{E}_\theta ||\widehat{\theta} - \theta||^2 &= \sum_{j=1}^n \left[\left(\frac{\sigma^2 \theta_j}{\sigma^2 + c^2}\right)^2 + \left(\frac{c^2}{c^2 + \sigma^2}\right)^2 \left(\frac{\sigma^2}{n}\right) \right] \\
&= \left(\frac{\sigma^2}{\sigma^2 + c^2}\right)^2 \sum_{j=1}^n \theta_j^2 + \sigma^2 \left(\frac{\sigma^2}{\sigma^2 + c^2}\right)^2 \\
&\leq c^2 \left(\frac{\sigma^2}{\sigma^2 + c^2}\right)^2 + \sigma^2 \left(\frac{\sigma^2}{\sigma^2 + c^2}\right)^2 \\
&= \frac{\sigma^2 c^2}{\sigma^2 + c^2}.
\end{aligned}
$$

Hence,

$$R_n \leq \frac{c^2 \sigma^2}{c^2 + \sigma^2}$$

for all n.

LOWER BOUND. Fix $0 < \delta < 1$. Let π_n be a Normal prior for which $\theta_1, \ldots, \theta_n$ are IID $N(0, c^2\delta^2/n)$. Let $B(\pi_n)$ denote the Bayes risk. Recall that $B(\pi_n)$ minimizes the integrated risk $B(\widehat{\theta}, \pi_n)$ over all estimators. The minimum is obtained by taking $\widehat{\theta}$ to be the posterior mean which has coordinates $\widehat{\theta}_j = c^2\delta^2 Z_j/(c^2\delta^2 + \sigma^2)$ with risk

$$R(\theta, \widehat{\theta}) = \sum_{i=1}^{n} \left[\theta_i^2 \left(\frac{\sigma_n^2}{\frac{c^2\delta^2}{n} + \sigma_n^2} \right)^2 + \sigma^2 \left(\frac{\frac{c^2\delta^2}{n}}{\frac{c^2\delta^2}{n} + \sigma_n^2} \right)^2 \right].$$

The Bayes risk is

$$B(\pi_n) = \int R(\theta, \widehat{\theta}) d\pi_n(\theta) = \frac{\sigma^2\delta^2 c^2}{\sigma^2 + \delta^2 c^2}.$$

So, for any estimator $\widehat{\theta}$,

$$
\begin{aligned}
B(\pi_n) & \leq B(\widehat{\theta}, \pi_n) \\
& = \int_{\Theta_n} R(\theta, \widehat{\theta}) d\pi_n + \int_{\Theta_n^c} R(\theta, \widehat{\theta}) d\pi_n \\
& \leq \sup_{\theta \in \Theta_n} R(\theta, \widehat{\theta}) + \int_{\Theta_n^c} R(\theta, \widehat{\theta}) d\pi_n \\
& \leq \sup_{\theta \in \Theta_n} R(\theta, \widehat{\theta}) + \sup_{\widehat{\theta}} \int_{\Theta_n^c} R(\theta, \widehat{\theta}) d\pi_n.
\end{aligned}
$$

Taking the infimum over all estimators that take values in Θ_n yields

$$B(\pi_n) \leq R_n + \sup_{\widehat{\theta}} \int_{\Theta_n^c} R(\theta, \widehat{\theta}) d\pi_n.$$

Hence,

$$
\begin{aligned}
R_n & \geq B(\pi_n) - \sup_{\widehat{\theta}} \int_{\Theta_n^c} R(\theta, \widehat{\theta}) d\pi_n \\
& = \frac{\sigma^2\delta^2 c^2}{\delta^2 c^2 + \sigma^2} - \sup_{\widehat{\theta}} \int_{\Theta_n^c} R(\theta, \widehat{\theta}) d\pi_n.
\end{aligned}
$$

Now, using the fact that $||a + b||^2 \leq 2(||a||^2 + ||b||^2)$, and the Cauchy–Schwartz inequality,

$$
\begin{aligned}
\sup_{\widehat{\theta}} \int_{\Theta_n^c} R(\theta, \widehat{\theta}) d\pi_n & \leq 2 \int_{\Theta_n^c} ||\theta||^2 d\pi_n + 2 \sup_{\widehat{\theta}} \int_{\Theta_n^c} \mathbb{E}_\theta ||\widehat{\theta}||^2 d\pi_n \\
& \leq 2\sqrt{\pi_n(\Theta_n^c)} \sqrt{\mathbb{E}_{\pi_n} \left(\sum_j \theta_j^2 \right)^2} + 2c^2 \pi_n(\Theta_n^c).
\end{aligned}
$$

Thus,

$$R_n \geq \frac{\sigma^2 \delta^2 c^2}{\sigma^2 + \delta^2 c^2} - 2\sqrt{\pi_n(\Theta_n^c)} \sqrt{\mathbb{E}_{\pi_n} \left(\sum_j \theta_j^2 \right)^2} - 2c^2 \pi_n(\Theta_n^c). \qquad (7.94)$$

We now bound the last two terms in (7.94).

We shall make use of the following large deviation inequality: if $Z_1, \ldots, Z_n \sim N(0,1)$ and $0 < t < 1$, then

$$\mathbb{P} \left(\left| \frac{1}{n} \sum_j (Z_j^2 - 1) \right| > t \right) \leq 2e^{-nt^2/8}.$$

Let $Z_j = \sqrt{n} \theta_j / (c\delta)$ and let $t = (1 - \delta^2)/\delta^2$. Then,

$$\begin{aligned}
\pi_n(\Theta_n^c) &= \mathbb{P} \left(\sum_{j=1}^n \theta_j^2 > c^2 \right) = \mathbb{P} \left(\frac{1}{n} \sum_{j=1}^n (Z_j^2 - 1) > t \right) \\
&\leq \mathbb{P} \left(\left| \frac{1}{n} \sum_j (Z_j^2 - 1) \right| > t \right) \leq 2e^{-nt^2/8}.
\end{aligned}$$

Next, we note that

$$\begin{aligned}
\mathbb{E}_{\pi_n} \left(\sum_j \theta_j^2 \right)^2 &= \sum_{i=1}^n \mathbb{E}_{\pi_n}(\theta_i^4) + \sum_{i=1}^n \sum_{j \neq i}^n \mathbb{E}_{\pi_n}(\theta_i^2) \mathbb{E}_{\pi_n}(\theta_j^2) \\
&= \frac{c^4 \delta^4 \mathbb{E}(Z_1^4)}{n} + \binom{n}{2} \frac{c^4 \delta^4}{n^2} = O(1).
\end{aligned}$$

Therefore, from (7.94),

$$R_n \geq \frac{\sigma^2 \delta^2 c^2}{\sigma^2 + \delta^2 c^2} - 2\sqrt{2} e^{-nt^2/16} O(1) - 2c^2 e^{-nt^2/8}.$$

Hence,

$$\liminf_{n \to \infty} R_n \geq \frac{\sigma^2 \delta^2 c^2}{\sigma^2 + \delta^2 c^2}.$$

The conclusion follows by letting $\delta \uparrow 1$. $\blacksquare$

Proof of Theorem 7.74. Let

$$a = \frac{\sigma_n}{n^{1/4}} (\log(1 + \epsilon^2))^{1/4}$$

and define

$$\Omega = \left\{ \theta = (\theta_1, \ldots, \theta_n) : |\theta_i| = a, \ i = 1, \ldots, n \right\}.$$

Note that Ω contains 2^n elements. Let f_θ denote the density of a multivariate Normal with mean θ and covariance $\sigma_n^2 I$ where I is the identity matrix. Define the mixture

$$q(y) = \frac{1}{2^n} \sum_{\theta \in \Omega} f_\theta(y).$$

Let f_0 denote the density of a multivariate Normal with mean $(0, \ldots, 0)$ and covariance $\sigma_n^2 I$. Then,

$$
\begin{aligned}
\int |f_0(x) - g(x)| dx &= \int \frac{|f_0(x) - g(x)|}{\sqrt{f_0(x)}} \sqrt{f_0(x)} dx \\
&\leq \sqrt{\int \frac{(f_0(x) - g(x))^2}{f_0(x)} dx} \\
&= \sqrt{\int \frac{g^2(x)}{f_0(x)} dx - 1}.
\end{aligned}
$$

Now,

$$
\begin{aligned}
\int \frac{q^2(x)}{f_0(x)} dx &= \int \left(\frac{q(x)}{f_0(x)} \right)^2 f_0(x) dx = \mathbb{E}_0 \left(\frac{q(x)}{f_0(x)} \right)^2 \\
&= \left(\frac{1}{2^n} \right)^2 \sum_{\theta, \nu \in \Omega} \mathbb{E}_0 \left(\frac{f_\theta(x) f_\nu(x)}{f_0^2(x)} \right) \\
&= \left(\frac{1}{2^n} \right)^2 \sum_{\theta, \nu \in \Omega} \exp \left\{ -\frac{1}{2\sigma_n^2} (\|\theta\|^2 + \|\nu\|^2) \right\} \mathbb{E}_0 \left(\exp \left\{ \epsilon^T (\theta + \nu) / \sigma_n^2 \right\} \right) \\
&= \left(\frac{1}{2^n} \right)^2 \sum_{\theta, \nu \in \Omega} \exp \left\{ -\frac{1}{2\sigma_n^2} (\|\theta\|^2 + \|\nu\|^2) \right\} \exp \left\{ \sum_{i=1}^n (\theta_i + \nu_i)^2 / (2\sigma_n^2) \right\} \\
&= \left(\frac{1}{2^n} \right)^2 \sum_{\theta, \nu \in \Omega} \exp \left\{ \frac{\langle \theta, \nu \rangle}{\sigma_n^2} \right\}.
\end{aligned}
$$

The latter is equal to the mean of $\exp(\langle \theta, \nu \rangle / \sigma_n^2)$ when drawing two vectors θ and ν at random from Ω. And this, in turn, is equal to

$$\mathbb{E} \exp \left\{ \frac{a^2 \sum_{i=1}^n E_i}{\sigma_n^2} \right\}$$

where $E_1, \ldots, E_n$ are independent and $\mathbb{P}(E_i = 1) = \mathbb{P}(E_i = -1) = 1/2$. Moreover,

$$
\begin{aligned}
\mathbb{E} \exp\left\{ \frac{a^2 \sum_{i=1}^{n} E_i}{\sigma_n^2} \right\} &= \prod_{i=1}^{n} \mathbb{E} \exp\left\{ \frac{a^2 E_i}{\sigma_n^2} \right\} \\
&= \left(\mathbb{E} \exp\left\{ \frac{a^2 E_1}{\sigma_n^2} \right\} \right)^n \\
&= \left(\cosh\left(\frac{a^2}{\sigma_n^2} \right) \right)^n
\end{aligned}
$$

where $\cosh(y) = (e^y + e^{-y})/2$. Thus,

$$
\int \frac{q^2(x)}{f_0(x)} dx = \left(\cosh\left(\frac{a^2}{\sigma_n^2} \right) \right)^n \leq e^{a^4 n / \sigma_n^4}
$$

where we have used the fact that $\cosh(y) \leq e^{y^2}$. Thus,

$$
\int |f_0(x) - q(x)| dx \leq \sqrt{e^{a^4 n / \sigma_n^4} - 1} = \epsilon.
$$

So, if Q denotes the probability measure with density q, we have, for any event A,

$$
\begin{aligned}
Q(A) &= \int_A q(x) dx = \int_A f_0(x) dx + \int_A (q(x) - f_0(x)) dx \\
&\geq \mathbb{P}_0(A) - \int_A |q(x) - f_0(x)| dx \geq \mathbb{P}_0(A) - \epsilon. \quad (7.95)
\end{aligned}
$$

Define two events, $A = \{(0, \ldots, 0) \in \mathcal{B}_n\}$ and $B = \{\Omega \cap \mathcal{B}_n \neq \emptyset\}$. Every $\theta \in \Omega$ has norm

$$
\|\theta\| = \sqrt{na^2} = \sigma_n n^{1/4} (\log(1 + \epsilon^2))^{1/4} \equiv c_n.
$$

Hence, $A \cap B \subset \{s_n \geq c_n\}$. Since $\mathbb{P}_\theta(\theta \in \mathcal{B}_n) \geq 1 - \alpha$ for all θ, it follows that $\mathbb{P}_\theta(B) \geq 1 - \alpha$ for all $\theta \in \Omega$. Hence, $Q(B) \geq 1 - \alpha$. From (7.95),

$$
\begin{aligned}
\mathbb{P}_0(s_n \geq c_n) &\geq \mathbb{P}_0(A \cap B) \geq Q(A \cap B) - \epsilon \\
&= Q(A) + Q(B) - Q(A \cup B) - \epsilon \\
&\geq Q(A) + Q(B) - 1 - \epsilon \\
&\geq Q(A) + (1 - \alpha) - 1 - \epsilon \\
&\geq \mathbb{P}_0(A) + (1 - \alpha) - 1 - 2\epsilon \\
&\geq (1 - \alpha) + (1 - \alpha) - 1 - 2\epsilon \\
&= 1 - 2\alpha - 2\epsilon.
\end{aligned}
$$

So, $\mathbb{E}_0(s_n) \geq (1 - 2\alpha - 2\epsilon)c_n$. It is easy to see that the same argument can be used for any $\theta \in \mathbb{R}^n$ and hence $\mathbb{E}_\theta(s_n) \geq (1 - 2\alpha - 2\epsilon)c_n$ for every $\theta \in \mathbb{R}^n$. ∎

Proof of Theorem 7.77. Let $a = \sigma_n z_{\alpha + 2\epsilon}$ where $0 < \epsilon < (1/2)(1/2 - \alpha)$ and define

$$\Omega = \left\{ \theta = (\theta_1, \ldots, \theta_n) : |\theta_i| = a, \ i = 1, \ldots, n \right\}.$$

Define the loss function $L = L(\widehat{\theta}, \theta) = \sum_{i=1}^n I(|\widehat{\theta}_i - \theta_i| \geq a)$. Let π be the uniform prior on Ω. The posterior mass function over Ω is $p(\theta|y) = \prod_{i=1}^n p(\theta_i|y_i)$ where

$$p(\theta_i|y_i) = \frac{e^{2ay_i/\sigma_n^2}}{1 + e^{2ay_i/\sigma_n^2}} I(\theta_i = a) + \frac{1}{1 + e^{2ay_i/\sigma_n^2}} I(\theta_i = -a).$$

The posterior risk is

$$\mathbb{E}(L(\widehat{\theta}, \theta)|y) = \sum_{i=1}^n \mathbb{P}(|\widehat{\theta}_i - \theta_i| \geq a|y_i)$$

which is minimized by taking $\widehat{\theta}_i = a$ if $y_i \geq 0$ and $\widehat{\theta}_i = -a$ if $y_i < 0$. The risk of this estimator is

$$\sum_{i=1}^n \left(\mathbb{P}(Y_i < 0|\theta_i = a)I(\theta_i = a) + \mathbb{P}(Y_i > 0|\theta_i = -a)I(\theta_i = -a) \right)$$
$$= n\Phi(-a/\sigma_n) = n(\alpha + 2\epsilon).$$

Since this risk is constant, it is the minimax risk. Therefore,

$$\inf_{\widehat{\theta}} \sup_{\theta \in \mathbb{R}^n} \sum_{i=1}^n \mathbb{P}_\theta(|\widehat{\theta}_i - \theta_i| \geq a) \geq \inf_{\widehat{\theta}} \sup_{\theta \in \Omega} \sum_{i=1}^n \mathbb{P}_\theta(|\widehat{\theta}_i - \theta_i| \geq a)$$
$$= n(\alpha + 2\epsilon).$$

Let $\gamma = \epsilon/(1 - \alpha - \epsilon)$. Given any estimator $\widehat{\theta}$,

$$\gamma n \mathbb{P}_\theta(L < \gamma n) + n \mathbb{P}_\theta(L \geq \gamma n) \geq L$$

and so

$$\sup_\theta \left(\gamma n \mathbb{P}_\theta(L < \gamma n) + n \mathbb{P}_\theta(L \geq \gamma n) \right) \geq \sup_\theta \mathbb{E}_\theta(L) \geq n(\alpha + 2\epsilon).$$

This inequality, together with the fact that $\mathbb{P}_\theta(L < \gamma n) + \mathbb{P}_\theta(L \geq \gamma n) = 1$ implies that

$$\sup_\theta \mathbb{P}_\theta(L \geq \gamma n) \geq \alpha + \epsilon.$$

Thus,

$$\sup_{\theta} \mathbb{P}_{\theta}(||\widehat{\theta} - \theta||^2 \geq \gamma n a^2) \geq \sup_{\theta} \mathbb{P}_{\theta}(L \geq \gamma n) \geq \alpha + \epsilon.$$

Therefore,

$$
\begin{aligned}
\sup_{\theta} \mathbb{P}_{\theta}(s_n^2 \geq \gamma n a^2) \quad &\geq \quad \sup_{\theta} \mathbb{P}_{\theta}(s_n^2 \geq ||\widehat{\theta} - \theta||^2 \geq \gamma n a^2) \\
&= \quad \sup_{\theta} \mathbb{P}_{\theta}(s_n^2 \geq ||\widehat{\theta} - \theta||^2) + \sup_{\theta} \mathbb{P}_{\theta}(||\widehat{\theta} - \theta||^2 \geq \gamma n a^2) - 1 \\
&\geq \quad \alpha + \epsilon + 1 - \alpha - 1 = \epsilon.
\end{aligned}
$$

Thus, $\sup_{\theta} E_{\theta}(s_n) \geq \epsilon a \sqrt{\gamma n}$. ∎

7.14   Exercises

1. Let $\theta_i = 1/i^2$ for $i = 1, \ldots, n$. Take $n = 1000$. Let $Z_i \sim N(\theta_i, 1)$ for $i = 1, \ldots, n$. Compute the risk of the MLE. Compute the risk of the estimator $\widetilde{\theta} = (bZ_1, bZ_2, \ldots, bZ_n)$. Plot this risk as a function of b. Find the optimal value b_*. Now conduct a simulation. For each run of the simulation, find the (modified) James–Stein estimator $\widehat{b}Z$ where

$$\widehat{b} = \left[1 - \frac{n}{\sum_i Z_i^2} \right]^{+}.$$

You will get one $\widehat{b}$ for each simulation. Compare the simulated values of $\widehat{b}$ to b_*. Also, compare the risk of the MLE and the James–Stein estimator (the latter obtained by simulation) to the Pinsker bound.

2. For the Normal means problem, consider the following *curved soft thresh-old estimator:*

$$
\widehat{\theta}_i = \begin{cases}
-(Z_i + \lambda)^2 & Z_i < -\lambda \\
0 & -\lambda \leq Z_i \leq \lambda \\
(Z_i - \lambda)^2 & Z_i > \lambda
\end{cases}
$$

where $\lambda > 0$ is some fixed constant.

(a) Find the risk of this estimator. *Hint:* $R = \mathbb{E}(\text{SURE})$.

(b) Consider problem (1). Use your estimator from (2a) with λ chosen from the data using SURE. Compare the risk to the risk of the James–Stein estimator. Now repeat the comparison for

$$\theta = (\overbrace{10, \ldots, 10}^{10 \text{ times}}, \overbrace{0, \ldots, 0}^{990 \text{ times}}).$$

3. Let $J = J_n$ be such that $J_n \to \infty$ and $n \to \infty$. Let

$$\widehat{\sigma}^2 = \frac{n}{J} \sum_{i=n-J+1}^{n} Z_i^2$$

where $Z_i \sim N(\theta_i, \sigma^2/n)$. Show that if $\theta = (\theta_1, \theta_2, \ldots)$ belongs to a Sobolev body of order $m > 1/2$ then $\widehat{\sigma}^2$ is a uniformly consistent estimator of σ^2 in the Normal means model.

4. Prove Stein's lemma: if $X \sim N(\mu, \sigma^2)$ then $\mathbb{E}(g(X)(X-\mu)) = \sigma^2 \mathbb{E} g'(X)$.

5. Verify equation 7.22.

6. Show that the hard threshold estimator defined in (7.23) is not weakly differentiable.

7. Compute the risk functions for the soft threshold estimator (7.21) and the hard threshold estimator (7.23).

8. Generate $Z_i \sim N(\theta_i, 1)$, $i = 1, \ldots, 100$, where $\theta_i = 1/i$. Compute a 95 percent confidence ball using: (i) the χ^2 confidence ball, (ii) the Baraud method, (iii) the pivotal method. Repeat 1000 times and compare the radii of the balls.

9. Let $||a - b||_\infty = \sup_j |a_j - b_j|$. Construct a confidence set B_n of the form $B_n = \{\theta \in \mathbb{R}^n : ||\theta - Z^n||_\infty \leq c_n\}$ such that $\mathbb{P}_\theta(\theta \in B_n) \geq 1 - \alpha$ for all $\theta \in \mathbb{R}^n$ under model (7.1) with $\sigma_n = \sigma/\sqrt{n}$. Find the expected diameter of your confidence set.

10. Consider Example 7.24. Define

$$\delta = \max_{S \in \mathcal{S}} \sup_{\theta \in \mathbb{R}^n} |\widehat{R}_S - R(\widehat{\theta}_S, \theta)|.$$

Try to bound δ in the following three cases: (i) $\mathcal{S}$ consists of a single model S; (ii) nested model selection; (iii) all subsets selection.

11. Consider Example 7.24. Another method for choosing a model is to use penalized likelihood. In particular, some well-known penalization model selection methods are **AIC** (Akaike's Information Criterion), Akaike (1973), **Mallows'** C_p, Mallows (1973), and **BIC** (Bayesian Information Criterion), Schwarz (1978). In the Normal means model, minimizing

SURE, AIC and C_p are equivalent. But BIC leads to a different model selection procedure. Specifically,

$$\mathrm{BIC}_B = \ell_B - \frac{|B|}{2} \log n$$

where ℓ_B is the log-likelihood of the submodel B evaluated at its maximum likelihood estimator. Find an explicit expression for BIC_B. Suppose we choose B by maximizing BIC_B over $\mathcal{B}$. Investigate the properties of this model selection procedure and compare it to selecting a model by minimizing SURE. In particular, compare the risk of the resulting estimators. Also, assuming there is a "true" submodel (that is, $\theta_i \neq 0$ if and only if $i \in B$), compare the probability of selecting the true submodel under each procedure. In general, estimating θ accurately and finding the true submodel are not the same. See Wasserman (2000).

12. By approximating the noncentral χ^2 with a Normal, find a large sample approximation to for ρ_0 and ρ_n in Example 7.69. Then prove equation (7.70).

8
Nonparametric Inference Using Orthogonal Functions

8.1 Introduction

In this chapter we use orthogonal function methods for nonparametric inference. Specifically, we use an orthogonal basis to convert regression and density estimation into a Normal means problem and then we construct estimates and confidence sets using the theory from Chapter 7. In the regression case, the resulting estimators are linear smoothers and thus are a special case of the estimators described in Section 5.2. We discuss another approach to orthogonal function regression based on wavelets in the next chapter.

8.2 Nonparametric Regression

The particular version of orthogonal function regression that we consider here was developed by Beran (2000) and Beran and Dümbgen (1998). They call the method REACT, which stands for **Risk Estimation and Adaptation after Coordinate Transformation**. Similar ideas have been developed by Efromovich (1999). In fact, the basic idea is quite old; see Cenčov (1962), for example.

Suppose we observe

$$Y_i = r(x_i) + \sigma\epsilon_i \tag{8.1}$$

where $\epsilon_i \sim N(0,1)$ are IID. For now, we assume a **regular design** meaning that $x_i = i/n$, $i = 1, \ldots, n$.

Let $\phi_1, \phi_2, \ldots$ be an orthonormal basis for $[0,1]$. In our examples we will often use the cosine basis:

$$\phi_1(x) \equiv 1, \qquad \phi_j(x) = \sqrt{2}\cos((j-1)\pi x), \ j \geq 2. \tag{8.2}$$

Expand r as

$$r(x) = \sum_{j=1}^{\infty} \theta_j \phi_j(x) \tag{8.3}$$

where $\theta_j = \int_0^1 \phi_j(x)r(x)dx$.

First, we approximate r by

$$r_n(x) = \sum_{j=1}^{n} \theta_j \phi_j(x)$$

which is the projection of r onto the span of $\{\phi_1, \ldots, \phi_n\}$.[1] This introduces an integrated squared bias of size

$$B_n(\theta) = \int_0^1 (r(x) - r_n(x))^2 dx = \sum_{j=n+1}^{\infty} \theta_j^2.$$

If r is smooth, this bias is quite small.

8.4 Lemma. *Let $\Theta(m,c)$ be a Sobolev ellipsoid.[2] Then,*

$$\sup_{\theta \in \Theta(m,c)} B_n(\theta) = O\left(\frac{1}{n^{2m}}\right). \tag{8.5}$$

In particular, if $m > 1/2$ then $\sup_{\theta \in \Theta(m,c)} B_n(\theta) = o(1/n)$.

Hence this bias is negligible and we shall ignore it for the rest of the chapter. More precisely, we will focus on estimating r_n rather than r. Our next task is to estimate the $\theta = (\theta_1, \ldots, \theta_n)$. Let

$$Z_j = \frac{1}{n}\sum_{i=1}^{n} Y_i\,\phi_j(x_i), \ j = 1, \ldots. \tag{8.6}$$

---

[1] More generally we could take $r_n(x) = \sum_{j=1}^{p(n)} \theta_j \phi_j(x)$ where $p(n) \to \infty$ at an appropriate rate.

[2] See Definition 7.2.

As we saw in equation (7.15) of Chapter 7,

$$Z_j \approx N\left(\theta_j, \frac{\sigma^2}{n}\right). \tag{8.7}$$

We know from the previous chapter that the MLE $Z = (Z_1, \ldots, Z_n)$ has large risk. One possibility for improving on the MLE is to use the James Stein estimator $\widehat{\theta}_{JS}$ defined in (7.41). We can think of the James–Stein estimator as the estimator that minimizes the estimated risk over all estimators of the form $(bZ_1, \ldots, bZ_n)$. REACT generalizes this idea by minimizing the risk over a larger class of estimators, called modulators.

A **modulator** is a vector $b = (b_1, \ldots, b_n)$ such that $0 \le b_j \le 1, j = 1, \ldots, n$. A **modulation estimator** is an estimator of the form

$$\widehat{\theta} = bZ = (b_1 Z_1, b_2 Z_2, \ldots, b_n Z_n). \tag{8.8}$$

A **constant modulator** is a modulator of the form $(b, \ldots, b)$. A **nested subset selection modulator** is a modulator of the form

$$b = (1, \ldots, 1, 0, \ldots, 0).$$

A **monotone modulator** is a modulator of the form

$$1 \ge b_1 \ge \cdots \ge b_n \ge 0.$$

The set of constant modulators is denoted by $\mathcal{M}_{\text{CONS}}$, the set of nested subset modulators is denoted by $\mathcal{M}_{\text{NSS}}$ and the set of monotone modulators is denoted by $\mathcal{M}_{\text{MON}}$.

Given a modulator $b = (b_1, \ldots, b_n)$, the function estimator is

$$\widehat{r}_n(x) = \sum_{j=1}^{n} \widehat{\theta}_j \phi_j(x) = \sum_{j=1}^{n} b_j Z_j \phi_j(x). \tag{8.9}$$

Observe that

$$\widehat{r}_n(x) = \sum_{i=1}^{n} Y_i \, \ell_i(x) \tag{8.10}$$

where

$$\ell_i(x) = \frac{1}{n} \sum_{j=1}^{n} b_j \phi_j(x) \phi_j(x_i). \tag{8.11}$$

Hence, $\widehat{r}_n$ is a linear smoother as described in Section 5.2.

Modulators shrink the Z_js towards 0 and, as we saw in the last chapter, shrinking tends to smooth the function. Thus, choosing the amount of shrinkage corresponds to the problem of choosing a bandwidth that we faced in

Chapter 5. We shall address the problem using Stein's unbiased risk estimator (Section 7.4) instead of cross-validation.

Let

$$R(b) = \mathbb{E}_\theta \left[\sum_{j=1}^n (b_j Z_j - \theta_j)^2 \right]$$

denote the risk of the estimator $\widehat{\theta} = (b_1 Z_1, \ldots, b_n Z_n)$. The idea of REACT is to estimate the risk $R(b)$ and choose $\widehat{b}$ to minimize the estimated risk over a class of modulators $\mathcal{M}$. Minimizing over $\mathcal{M}_{\text{CONS}}$ yields the James–Stein estimator, so REACT is a generalization of James–Stein estimation.

To proceed, we need to estimate σ. Any of the methods discussed in Chapter 5 can be used. Another estimator, well-suited for the present framework, is

$$\widehat{\sigma}^2 = \frac{1}{n - J_n} \sum_{i=n-J_n+1}^n Z_i^2. \tag{8.12}$$

This estimator is consistent as long as $J_n \to \infty$ and $n - J_n \to \infty$ as $n \to \infty$. As a default value, $J_n = n/4$ is not unreasonable. The intuition is that if r is smooth then we expect $\theta_j \approx 0$ for large j, and hence $Z_j^2 = (\theta_j + \sigma \epsilon_j/\sqrt{n})^2 \approx (\sigma \epsilon_j/\sqrt{n})^2 = \sigma^2 \epsilon_j^2/n$. Therefore,

$$\mathbb{E}(\widehat{\sigma}^2) = \frac{1}{n - J_n} \sum_{n - J_n + 1}^n \mathbb{E}(Z_i^2) \approx \frac{1}{n - J_n} \sum_{n - J_n + 1}^n \frac{\sigma^2}{n} \mathbb{E}(\epsilon_i^2) = \sigma^2.$$

Now we can estimate the risk function.

8.13 Theorem. *The risk of a modulator b is*

$$R(b) = \sum_{j=1}^n \theta_j^2 (1 - b_j)^2 + \frac{\sigma^2}{n} \sum_{j=1}^n b_j^2. \tag{8.14}$$

The (modified)[3] SURE estimator of $R(b)$ is

$$\widehat{R}(b) = \sum_{j=1}^n \left(Z_j^2 - \frac{\widehat{\sigma}^2}{n} \right)_+ (1 - b_j)^2 + \frac{\widehat{\sigma}^2}{n} \sum_{j=1}^n b_j^2 \tag{8.15}$$

where $\widehat{\sigma}^2$ is a consistent estimate of σ^2 such as (8.12).

---

[3]We call this a modified risk estimator since we have inserted an estimate $\widehat{\sigma}$ of σ and we replaced $(Z_j^2 - \widehat{\sigma}^2/n)$ with $(Z_j^2 - \widehat{\sigma}^2/n)_+$ which usually improves the risk estimate.

8.16 Definition. *Let* $\mathcal{M}$ *be a set of modulators. The* **modulation estimator** *of* θ *is* $\widehat{\theta} = (\widehat{b}_1 Z_1, \ldots, \widehat{b}_n Z_n)$ *where* $\widehat{b} = (\widehat{b}_1 \ldots, \widehat{b}_n)$ *minimizes* $\widehat{R}(b)$ *over* $\mathcal{M}$. *The* REACT **function estimator** *is*

$$\widehat{r}_n(x) = \sum_{j=1}^{n} \widehat{\theta}_j \phi_j(x) = \sum_{j=1}^{n} \widehat{b}_j Z_j \phi_j(x).$$

For a fixed b, we expect that $\widehat{R}(b)$ approximates $R(b)$. But for the REACT estimator we require more: we want $\widehat{R}(b)$ to approximate $R(b)$ uniformly for $b \in \mathcal{M}$. If so, then $\inf_{b \in \mathcal{M}} \widehat{R}(b) \approx \inf_{b \in \mathcal{M}} R(b)$ and the b that minimizes $\widehat{R}(b)$ should be nearly as good as the b that minimizes $R(b)$. This motivates the next result.

8.17 Theorem (Beran and Dümbgen, 1998). *Let* $\mathcal{M}$ *be one of* $\mathcal{M}_{\text{CONS}}$, $\mathcal{M}_{\text{NSS}}$ *or* $\mathcal{M}_{\text{MON}}$. *Let* $R(b)$ *denote the true risk of the estimator* $(b_1 Z_1, \ldots, b_n Z_n)$. *Let* b^* *minimize* $R(b)$ *over* $\mathcal{M}$ *and let* $\widehat{b}$ *minimize* $\widehat{R}(b)$ *over* $\mathcal{M}$. *Then*

$$|R(\widehat{b}) - R(b^*)| \to 0$$

as $n \to \infty$. *For* $\mathcal{M} = \mathcal{M}_{\text{CONS}}$ *or* $\mathcal{M} = \mathcal{M}_{\text{MON}}$, *the estimator* $\widehat{\theta} = (\widehat{b}_1 Z_1, \ldots, \widehat{b}_n Z_n)$ *achieves the Pinsker bound (7.29).*

To implement this method, we need to find $\widehat{b}$ to minimize $\widehat{R}(b)$. The minimum of $\widehat{R}(b)$ over $\mathcal{M}_{\text{CONS}}$ is the James–Stein estimator. To minimize $\widehat{R}(b)$ over $\mathcal{M}_{\text{NSS}}$, we compute $\widehat{R}(b)$ for every modulator of the form $(1, 1 \ldots, 1, 0, \ldots, 0)$ and then the minimum is found. In other words, we find $\widehat{J}$ to minimize

$$\widehat{R}(J) = \frac{J\widehat{\sigma}^2}{n} + \sum_{j=J+1}^{n} \left(Z_j^2 - \frac{\widehat{\sigma}^2}{n} \right)_{+} \tag{8.18}$$

and set $\widehat{r}_n(x) = \sum_{j=1}^{\widehat{J}} Z_j \phi_j(x)$. It is a good idea to plot the estimated risk as a function of J. To minimize $\widehat{R}(b)$ over $\mathcal{M}_{\text{MON}}$, note that $\widehat{R}(b)$ can be written as

$$\widehat{R}(b) = \sum_{i=1}^{n} (b_i - g_i)^2 Z_i^2 + \frac{\widehat{\sigma}^2}{n} \sum_{i=1}^{n} g_i \tag{8.19}$$

where $g_i = (Z_i^2 - (\widehat{\sigma}^2/n))/Z_i^2$. So it suffices to minimize

$$\sum_{i=1}^{n} (b_i - g_i)^2 Z_i^2$$

subject to $b_1 \geq \cdots \geq b_n$. This is simply a weighted least squares problem subject to a monotonicity constraint. There is a well-known algorithm called the pooled-adjacent-violators (PAV) algorithm for performing this minimization; See Robertson et al. (1988).

Usually, monotone modulators lead to estimates that are close to the NSS modulators and the latter are very easy to implement. Thus, as a default, the NSS method is reasonable. At this point, we can summarize the whole REACT procedure.

Summary of REACT

1. Let $Z_j = n^{-1} \sum_{i=1}^{n} Y_i \, \phi_j(x_i)$ for $j = 1, \ldots, n$.

2. Find $\widehat{J}$ to minimize the risk estimator $\widehat{R}(J)$ given by equation (8.18).

3. Let

$$\widehat{r}_n(x) = \sum_{j=1}^{\widehat{J}} Z_j \phi_j(x).$$

8.20 Example (Doppler function). Recall that the Doppler function from Example 5.63 is

$$r(x) = \sqrt{x(1-x)} \sin\left(\frac{2.1\pi}{x + .05}\right).$$

The top left panel in Figure 8.1 shows the true function. The top right panel shows 1000 data points. The data were simulated from the model $Y_i = r(i/n) + \sigma \epsilon_i$ with $\sigma = 0.1$ and $\epsilon_i \sim N(0, 1)$. The bottom left panel shows the estimated risk for the NSS modulator as a function of the number of terms in the fit. The risk was minimized by using the modulator:

$$b = (\underbrace{1, \ldots, 1}_{187}, \underbrace{0, \ldots, 0}_{813}).$$

The bottom right panel shows the REACT fit. Compare with Figure 5.6. ∎

8.21 Example (CMB data). Let us compare REACT to local smoothing for the CMB data from Example 4.4. The estimated risk (for NSS) is minimized by using $J = 6$ basis functions. The fit is shown in Figure 8.2. and is similar to the fits obtained in Chapter 5. (We are ignoring the fact that the variance is not constant.) The plot of the risk reveals that there is another local minimum around $J = 40$. The bottom right plot shows the fit using 40 basis functions. This fit appears to undersmooth. ∎

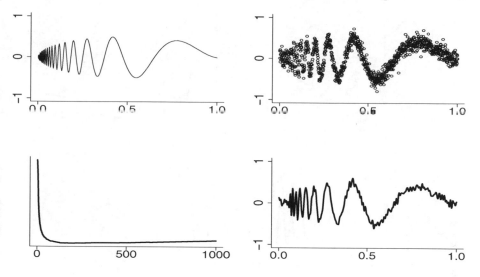

FIGURE 8.1. Doppler test function. Top left: true function. Top right: 1000 data points. Bottom left: estimated risk as a function of the number of terms in the fit. Bottom right: final REACT fit.

There are several ways to construct a confidence set for r. We begin with confidence balls. First, construct a confidence ball $\mathcal{B}_n$ for $\theta = (\theta_1, \ldots, \theta_n)$ using any of the methods in Section 7.8. Then define

$$\mathcal{C}_n = \left\{ r = \sum_{j=1}^{n} \theta_j \phi_j : (\theta_1, \ldots, \theta_n) \in \mathcal{B}_n \right\}. \tag{8.22}$$

It follows that $\mathcal{C}_n$ is a confidence ball for r_n. If we use the pivotal method from Section 7.8 we get the following.

8.23 Theorem (Beran and Dümbgen, 1998). *Let $\widehat{\theta}$ be the MON or NSS estimator and let $\widehat{\sigma}^2$ be the estimator of σ defined in (8.12). Let*

$$\mathcal{B}_n = \left\{ \theta = (\theta_1, \ldots, \theta_n) : \sum_{j=1}^{n} (\theta_j - \widehat{\theta}_j)^2 \leq s_n^2 \right\} \tag{8.24}$$

where

$$s_n^2 = \widehat{R}(\widehat{b}) + \frac{\widehat{\tau} z_\alpha}{\sqrt{n}}$$

$$\widehat{\tau}^2 = \frac{2\widehat{\sigma}^4}{n} \sum_j \left((2\widehat{b}_j - 1)(1 - c_j) \right)^2$$

$$+ 4\widehat{\sigma}^2 \sum_j \left(Z_j^2 - \frac{\widehat{\sigma}^2}{n} \right) \left((1 - \widehat{b}_j) + (2\widehat{b}_j - 1)c_j \right)^2$$

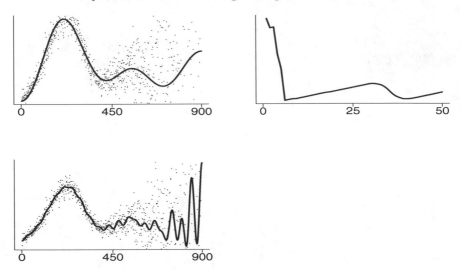

FIGURE 8.2. CMB data using REACT. Top left: NSS fit using $J = 6$ basis functions. Top right: estimated risk. Bottom left: NSS fit using $J = 40$ basis functions.

and

$$c_j = \begin{cases} 0 & \text{if } j \leq n - J \\ 1/J & \text{if } j > n - J. \end{cases}$$

Then, for any $c > 0$ and $m > 1/2$,

$$\lim_{n \to \infty} \sup_{\theta \in \Theta(m,c)} |\mathbb{P}_\theta(\theta \in \mathcal{B}_n) - (1 - \alpha)| = 0.$$

To construct confidence bands, we use the fact that $\hat{r}_n$ is a linear smoother and we can then use the method from Section 5.7. The band is given by (5.99), namely,

$$I(x) = \left(\hat{r}_n(x) - c\,\hat{\sigma}||\ell(x)||, \ \hat{r}_n(x) + c\,\hat{\sigma}||\ell(x)|| \right) \tag{8.25}$$

where

$$||\ell(x)||^2 \approx \frac{1}{n} \sum_{j=1}^{n} b_j^2 \phi_j^2(x) \tag{8.26}$$

and c is from equation (5.102).

8.3 Irregular Designs

So far we have assumed a regular design $x_i = i/n$. Now let us relax this assumption and deal with an **irregular design**. There are several ways to handle this case. The simplest is to use a basis $\{\phi_1, \ldots, \phi_n\}$ that is orthogonal with respect to the design points $x_1, \ldots, x_n$. That is, we choose a basis for $L_2(P_n)$ where $P_n = n^{-1} \sum_{i=1}^{n} \delta_i$ and δ_i is a point mass at x_i. This requires that

$$||\phi_j^2|| = 1, \quad j = 1, \ldots, n$$

and

$$\langle \phi_j, \phi_k \rangle = 0, \quad 1 \leq j < k \leq n$$

where

$$\langle f, g \rangle = \int f(x)g(x)dP_n(x) = \frac{1}{n} \sum_{i=1}^{n} f(x_i)g(x_i)$$

and

$$||f||^2 = \int f^2(x)dP_n(x) = \frac{1}{n} \sum_{i=1}^{n} f^2(x_i).$$

We can construct such a basis by Gram–Schmidt orthogonalization as follows. Let $g_1, \ldots, g_n$ be any convenient orthonormal basis for $\mathbb{R}^n$. Let

$$\phi_1(x) = \frac{\psi_1(x)}{||\psi_1||} \quad \text{where} \quad \psi_1(x) = g_1(x)$$

and for $2 \leq r \leq n$ define

$$\phi_r(x) = \frac{\psi_r(x)}{||\psi_r||} \quad \text{where} \quad \psi_r(x) = g_r(x) - \sum_{j=1}^{r-1} a_{r,j}\phi_j(x)$$

and

$$a_{r,j} = \langle g_r, \phi_j \rangle.$$

Then, $\phi_1, \ldots, \phi_n$ form an orthonormal basis with respect to P_n.

Now, as before, we define

$$Z_j = \frac{1}{n} \sum_{i=1}^{n} Y_i \, \phi_j(x_i), \quad j = 1, \ldots, n. \tag{8.27}$$

It follows that

$$Z_j \approx N\left(\theta_j, \frac{\sigma^2}{n}\right)$$

and we can then use the methods that we developed in this chapter.

8.4   Density Estimation

Orthogonal function methods can also be used for density estimation. Let $X_1, \ldots, X_n$ be an IID sample from a distribution F with density f with support on $(0,1)$. We assume that $f \in L_2(0,1)$ so we can expand f as

$$f(x) = \sum_{j=1}^{\infty} \theta_j \phi_j(x)$$

where, as before, $\phi_1, \phi_2, \ldots$, is an orthogonal basis. Let

$$Z_j = \frac{1}{n} \sum_{i=1}^{n} \phi_j(X_i), \quad j = 1, 2, \ldots, n. \tag{8.28}$$

Then,

$$\mathbb{E}(Z_j) = \int \phi_j(x) f(x) dx = \theta_j$$

and

$$\mathbb{V}(Z_j) = \frac{1}{n} \left(\int \phi_j^2(x) f(x) dx - \theta_j^2 \right) \equiv \sigma_j^2.$$

As in the regression case, we take $\widehat{\theta}_j = 0$ for $j > n$ and we estimate $\theta = (\theta_1, \ldots, \theta_n)$ using a modulation estimator $\widehat{\theta} = bZ = (b_1 Z_1, \ldots, b_n Z_n)$. The risk of this estimator is

$$R(b) = \sum_{j=1}^{n} b_j^2 \sigma_j^2 + \sum_{j=1}^{n} (1 - b_j)^2 \theta_j^2. \tag{8.29}$$

We estimate σ_j^2 by

$$\widehat{\sigma}_j^2 = \frac{1}{n^2} \sum_{i=1}^{n} (\phi_j(X_i) - Z_j)^2$$

and θ_j^2 by $Z_j^2 - \widehat{\sigma}_j^2$; then we can estimate the risk by

$$\widehat{R}(b) = \sum_{j=1}^{n} b_j^2 \widehat{\sigma}_j^2 + \sum_{j=1}^{n} (1 - b_j)^2 \left(Z_j^2 - \widehat{\sigma}_j^2 \right)_+ . \tag{8.30}$$

Finally, we choose $\widehat{b}$ by minimizing $\widehat{R}(b)$ over some class of modulators $\mathcal{M}$. The density estimate can be negative. We can fix this by performing surgery: remove the negative part of the density and renormalize it to integrate to 1. Better surgery methods are discussed in Glad et al. (2003).

8.5   Comparison of Methods

The methods we have introduced for nonparametric regression so far are local regression (Section 5.4), spline smoothing (Section 5.5), and orthogonal function smoothing. In many ways, these methods are very similar. They all involve a bias variance tradeoff and they all require choosing a smoothing parameter. Local polynomial smoothers have the advantage that they automatically correct for boundary bias. It is possible to modify orthogonal function estimators to alleviate boundary bias by changing the basis slightly; see Efromovich (1999). An advantage of orthogonal function smoothing is that it converts nonparametric regression into the many Normal means problem, which is simpler, at least for theoretical purposes. There are rarely huge differences between the approaches, especially when these differences are judged relative to the width of confidence bands. Each approach has its champions and its detractors. It is wise to use all available tools for each problem. If they agree then the choice of method is one of convenience or taste; if they disagree then there is value in figuring out why they differ.

Finally, let us mention that there is a formal relationship between these approaches. For example, orthogonal function can be viewed as kernel smoothing with a particular kernel and vice versa. See Härdle et al. (1998) for details.

8.6   Tensor Product Models

The methods in this chapter extend readily to higher dimensions although our previous remarks about the curse of dimensionality apply here as well.

Suppose that $r(x_1, x_2)$ is a function of two variables. For simplicity, assume that $0 \leq x_1, x_2 \leq 1$. If $\phi_0, \phi_1, \ldots$ is an orthonormal basis for $L_2(0, 1)$ then the functions

$$\left\{ \phi_{j,k}(x_1, x_2) = \phi_j(x_1)\phi_k(x_2) : \ j, k = 0, 1, \ldots, \right\}$$

form an orthonormal basis for $L_2([0, 1] \times [0, 1])$, called the **tensor product basis.** The basis can be extended to d dimensions in the obvious way.

Suppose that $\phi_0 = 1$. Then a function $r \in L_2([0, 1] \times [0, 1])$ can be expanded in the tensor product basis as

$$
\begin{aligned}
r(x_1, x_2) \ &= \ \sum_{j,k=0}^{\infty} \beta_{j,k}\, \phi_j(x_1)\phi_k(x_2) \\
&= \ \beta_0 + \sum_{j=1}^{\infty} \beta_{j,0}\, \phi_j(x_1) + \sum_{j=1}^{\infty} \beta_{0,j}\, \phi_j(x_2) + \sum_{j,k=1}^{\infty} \beta_{j,k}\, \phi_j(x_1)\phi_k(x_2).
\end{aligned}
$$

This expansion has an ANOVA-like structure consisting of a mean, main effects, and interactions. This structure suggests a way to get better estimators. We could put stronger smoothness assumptions on higher-order interactions to get better rates of convergence (at the expense of more assumptions). See Lin (2000), Wahba et al. (1995), Gao et al. (2001), and Lin et al. (2000).

8.7   Bibliographic Remarks

The REACT method is developed in Beran (2000) and Beran and Dümbgen (1998). A different approach to using orthogonal functions is discussed in Efromovich (1999). REACT confidence sets are extended to nonconstant variance in Genovese et al. (2004), to wavelets in Genovese and Wasserman (2005), and to density estimation in Jang et al. (2004).

8.8   Exercises

1. Prove Lemma 8.4.

2. Prove Theorem 8.13.

3. Prove equation (8.19).

4. Prove equation (8.26).

5. Show that the estimator (8.12) is consistent.

6. Show that the estimator (8.12) is uniformly consistent over Sobolev ellipsoids.

7. Get the data on fragments of glass collected in forensic work from the book website. Let Y be refractive index and let x be aluminium content (the fourth variable). Perform a nonparametric regression to fit the model $Y = r(x) + \epsilon$. Use REACT and compare to local linear smoothing. Estimate the variance. Construct 95 percent confidence bands for your estimate.

8. Get the motorcycle data from the book website. The covariate is time (in milliseconds) and the response is acceleration at time of impact. Use REACT to fit the data. Compute 95 percent confidence bands. Compute a 95 percent confidence ball. Can you think of a creative way to display the confidence ball?

9. Generate 1000 observations from the model $Y_i = r(x_i) + \sigma \epsilon_i$ where $x_i = i/n$, $\epsilon_i \sim N(0,1)$ and r is the Doppler function. Make three data sets corresponding to $\sigma = .1$, $\sigma = 1$ and $\sigma = 3$. Estimate the function using local linear regression and using REACT. In each case, compute a 95 percent confidence band. Compare the fits and the confidence bands.

10. Repeat the previous exercise but use Cauchy errors instead of Normal errors. How might you change the procedure to make the estimators more robust?

11. Generate $n = 1000$ data points from $(1/2)N(0,1) + (1/2)N(\mu, 1)$. Compare kernel density estimators and REACT density estimators. Try $\mu = 0, 1, 2, 3, 4, 5$.

12. Recall that a modulator is any vector of the form $b = (b_1, \ldots, b_n)$ such that $0 \le b_j \le 1$, $j = 1, \ldots, n$. The **greedy modulator** is the modulator $b^* = (b_1^*, \ldots, b_n^*)$ chosen to minimize the risk $R(b)$ over all modulators.

(a) Find b^*.

(b) What happens if we try to estimate b^* from the data? In particular, consider taking $\widehat{b}^*$ to minimize the estimated risk $\widehat{R}$. Why will this not work well? (The problem is that we are now minimizing $\widehat{R}$ over a very large class and $\widehat{R}$ does not approximate R uniformly over such a large class.)

13. Let
$$Y_i = r(x_{1i}, x_{2i}) + \epsilon_i$$
where $\epsilon_i \sim N(0,1)$, $x_{1i} = x_{2i} = i/n$ and $r(x_1, x_2) = x_1 + \cos(x_2)$. Generate 1000 observations. Fit a tensor product model with J_1 basis elements for x_1 and J_2 basis elements for x_2. Use SURE (Stein's unbiased risk estimator) to choose J_1 and J_2.

14. Download the air quality data set from the book website. Model ozone as a function of solar R, wind and temperature. Use a tensor product basis.

9
Wavelets and Other Adaptive Methods

This chapter concerns estimating functions that are **spatially inhomogeneous**, functions $r(x)$ whose smoothness varies substantially with x. For example, Figure 9.1 shows the "blocks" function whose definition is given in Example 9.39. The function is very smooth except for several abrupt jumps. The top right plot shows 100 data points drawn according to $Y_i = r(x_i) + \epsilon_i$ with $\epsilon_i \sim N(0, 1)$, and the x_is equally spaced.

It might be difficult to estimate r using the methods we have discussed so far. If we use local regression with a large bandwidth, then we will smooth out the jumps; if we use a small bandwidth, then we will find the jumps but we will make the rest of the curve very wiggly. If we use orthogonal functions and keep only low-order terms, we will miss the jumps; if we allow higher-order terms we will find the jumps but we will make the rest of the curve very wiggly. The function estimates in Figure 9.1 illustrate this point. Another example of an inhomogeneous function is the Doppler function in Example 5.63.

Estimators that are designed to estimate such functions are said to be **spatially adaptive** or **locally adaptive**. A closely related idea is to find globally adaptive estimators; these are function estimators that perform well over large classes of function spaces. In this chapter we explore adaptive estimators with an emphasis on **wavelet methods**. In Section 9.9 we briefly consider some other adaptive methods.

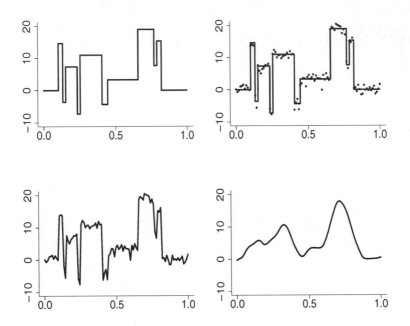

FIGURE 9.1. The blocks function (upper left plot) is inhomogeneous. One-hundred data points are shown in the upper right plot. A local linear smoother with a small bandwidth (lower left) picks up the jumps but adds many wiggles. A local linear smoother with a large bandwidth (lower right) is smooth but misses the jumps.

Caveat Emptor! Before proceeding, a warning is in order. Adaptive estimation is difficult. Unless the signal-to-noise ratio is large, we cannot expect to adapt very well. To quote from Loader (1999b):

> Locally adaptive procedures work well on examples with plenty of data, obvious structure and low noise. But ... these are not difficult problems .... The real challenges ...occur when the structure is not obvious, and there are questions as to which features in a dataset are real. In such cases, simpler methods ... are most useful, and locally adaptive methods produce little benefit.

The reader is urged to keep this in mind as we proceed. See Section 9.10 for more discussion on this point. Despite the caveat, the methods in this chapter are important because they do work well in high signal-to-noise problems and, more importantly, the conceptual ideas underlying the methods are important in their own right.

The wavelet methods discussed in this chapter illustrate a concept that is becoming increasingly important in statistics and machine learning, namely, the notion of **sparseness**. A function $f = \sum_j \beta_j \phi_j$ is **sparse** in a basis $\phi_1, \phi_2, \ldots$ if most of the β_j's are zero (or close to zero). We will see that even some fairly complicated functions are sparse when expanded in a wavelet basis. Sparseness generalizes smoothness: smooth functions are sparse but there are also nonsmooth functions that are sparse. It is interesting to note that sparseness is not captured well by the L_2 norm but it is captured by the L_1 norm. For example, consider the n-vectors $a = (1, 0, \ldots, 0)$ and $b = (1/\sqrt{n}, \ldots, 1/\sqrt{n})$. Then both have the same L_2 norm: $||a||_2 = ||b||_2 = 1$. However, the L_1 norms are $||a||_1 = \sum_i |a_i| = 1$ and $||b||_1 = \sum_i |b_i| = \sqrt{n}$. The L_1 norms reflects the sparsity of a.

NOTATION. Throughout this chapter, $\mathbb{Z}$ denotes the set of integers and $\mathbb{Z}_+$ denotes the set of positive integers. The **Fourier transform** f^* of a function f is

$$f^*(t) = \int_{-\infty}^{\infty} e^{-ixt} f(x) dx \qquad (9.1)$$

where $i = \sqrt{-1}$. If f^* is absolutely integrable then f can be recovered at almost all x by the **inverse Fourier transform**

$$f(x) = \frac{1}{2\pi} \int_{-\infty}^{\infty} e^{ixt} f^*(t) dt. \qquad (9.2)$$

Given any function f and integers j and k, define

$$f_{jk}(x) = 2^{j/2} f(2^j x - k). \qquad (9.3)$$

9.1   Haar Wavelets

We begin with a simple wavelet called the Haar wavelet. The **Haar father wavelet** or **Haar scaling function** is defined by

$$\phi(x) = \begin{cases} 1 & \text{if } 0 \leq x < 1 \\ 0 & \text{otherwise.} \end{cases} \qquad (9.4)$$

The **mother Haar wavelet** is defined by

$$\psi(x) = \begin{cases} -1 & \text{if } 0 \leq x \leq \frac{1}{2} \\ 1 & \text{if } \frac{1}{2} < x \leq 1. \end{cases} \qquad (9.5)$$

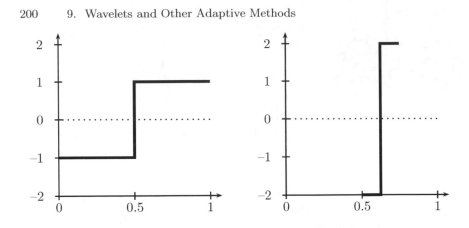

FIGURE 9.2. Some Haar wavelets. Left: the mother wavelet $\psi(x)$; right: $\psi_{2,2}(x)$.

For any integers j and k define $\phi_{jk}(x)$ and $\psi_{jk}(x)$ as in (9.3). The function ψ_{jk} has the same shape as ψ but it has been rescaled by a factor of $2^{j/2}$ and shifted by a factor of k. See Figure 9.2 for some examples of Haar wavelets.

Let

$$W_j = \{\psi_{jk}, \ k = 0, 1, \ldots, 2^j - 1\}$$

be the set of rescaled and shifted mother wavelets at resolution j.

9.6 Theorem. *The set of functions*

$$\{\phi, W_0, W_1, W_2, \ldots, \}$$

is an orthonormal basis for $L_2(0, 1)$.

It follows from this theorem that we can expand any function $f \in L_2(0, 1)$ in this basis. Because each W_j is itself a set of functions, we write the expansion as a double sum:

$$f(x) = \alpha\, \phi(x) + \sum_{j=0}^{\infty} \sum_{k=0}^{2^j - 1} \beta_{jk} \psi_{jk}(x) \tag{9.7}$$

where

$$\alpha = \int_0^1 f(x)\phi(x)\, dx, \quad \beta_{jk} = \int_0^1 f(x)\psi_{jk}(x)\, dx.$$

We call α the **scaling coefficient** and the β_{jk}'s are called the **detail coefficients**. We call the finite sum

$$f_J(x) = \alpha\phi(x) + \sum_{j=0}^{J-1} \sum_{k=0}^{2^j - 1} \beta_{jk} \psi_{jk}(x) \tag{9.8}$$

the **resolution** J approximation to f. The total number of terms in this sum is

$$1 + \sum_{j=0}^{J-1} 2^j = 1 + 2^J - 1 = 2^J.$$

9.9 Example. Figure 9.3 shows the Doppler signal (Example 5.63) and its resolution J approximation for $J = 3, 5$ and $J = 8$. ∎

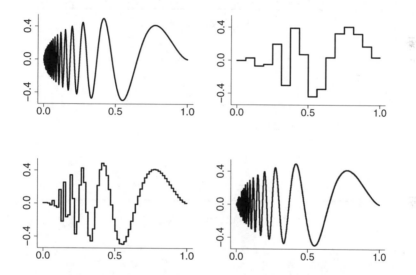

FIGURE 9.3. The Doppler signal (top left) and its reconstruction $f_J(x) = \alpha\phi(x) + \sum_{j=0}^{J-1} \sum_k \beta_{jk}\psi_{jk}(x)$ based on $J = 3$ (top right), $J = 5$ (bottom left) and $J = 8$ (bottom right).

When j is large, ψ_{jk} is a very localized function. This makes it possible to add a blip to a function in one place without adding wiggles elsewhere. This is what makes a wavelet basis a good tool for modeling inhomogeneous functions.

Figure 9.4 shows that blocks function and the coefficients of the expansion of the function in the Haar basis. Notice that the expansion is sparse (most coefficients are zero) since nonzero coefficients are needed mainly where the function has jumps.

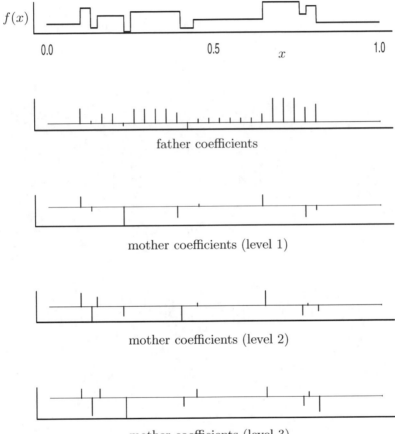

FIGURE 9.4. The blocks function $f(x)$ is shown on top. The second plot show the coefficients of the father wavelets. The third plot show the coefficients of the second-level mother wavelets. The remaining plots show the coefficients of the mother wavelets at higher levels. Despite the fact that the function is not smooth, the function is sparse: most coefficients are zero.

9.2   Constructing Wavelets

Haar wavelets are useful because they are localized, that is, they have bounded support. But Haar wavelets are not smooth. How can we construct other wavelets? In particular, how can we construct localized, smooth wavelets? The answer is quite involved. We will give a brief synopsis of the main ideas, see Härdle et al. (1998) and Daubechies (1992) for more details.

Given any function ϕ define

$$
\begin{aligned}
V_0 &= \left\{ f: \ f(x) = \sum_{k \in \mathbb{Z}} c_k \phi(x - k), \ \sum_{k \in \mathbb{Z}} c_k^2 < \infty \right\}, \\
V_1 &= \left\{ f(x) = g(2x): \ g \in V_0 \right\}, \\
V_2 &= \left\{ f(x) = g(2x): \ g \in V_1 \right\},
\end{aligned}
\qquad (9.10)
$$

$$\vdots \quad \vdots \quad \vdots$$

9.11 Definition. *Given a function ϕ, define $V_0, V_1, \ldots,$ as in (9.10). We say that ϕ generates a* **multiresolution analysis (MRA)** *of $\mathbb{R}$ if*

$$V_j \subset V_{j+1}, \quad j \geq 0, \qquad (9.12)$$

and

$$\bigcup_{j \geq 0} V_j \quad \text{is dense in } L_2(\mathbb{R}). \qquad (9.13)$$

We call ϕ the **father wavelet** *or the* **scaling function.**

Equation (9.13) means that for any function $f \in L_2(\mathbb{R})$, there exists a sequence of functions $f_1, f_2, \ldots$ such that each $f_r \in \bigcup_{j \geq 0} V_j$ and $||f_r - f|| \to 0$ as $r \to \infty$.

9.14 Lemma. *If $V_0, V_1, \ldots$ is an MRA generated by ϕ, then $\{\phi_{jk}, \ k \in \mathbb{Z}\}$ is an orthonormal basis for V_j.*

9.15 Example. If ϕ is the father Haar wavelet, then V_j is the set of functions $f \in L_2(\mathbb{R})$ that are piecewise constant on $[k2^{-j}, (k+1)2^{-j})$ for $k \in \mathbb{Z}$. It is easy to check that $V_0, V_1, \ldots$ form an MRA. ∎

Suppose we have an MRA. Since $\phi \in V_0$ and $V_0 \subset V_1$, we also have that $\phi \in V_1$. Since $\{\phi_{1k}, \ k \in \mathbb{Z}\}$ is an orthonormal basis for V_1, we can thus write

ϕ as a linear combination of functions in V_1:

$$\phi(x) = \sum_k \ell_k \, \phi_{1k}(x) \tag{9.16}$$

where $\ell_k = \int \phi(x)\phi_{1k}(x)dx$ and $\sum_k \ell_k^2 < \infty$. Equation (9.16) is called the **two-scale relationship** or **the dilation equation**. For Haar wavelets $\ell_0 = \ell_1 = 2^{-1/2}$ and $\ell_k = 0$ for $k \neq 0, 1$. The coefficients $\{\ell_k\}$ are called **scaling coefficients**. The two-scale relationship implies that

$$\phi^*(t) = m_0(t/2)\phi^*(t/2) \tag{9.17}$$

where $m_0(t) = \sum_k \ell_k e^{-itk}/\sqrt{2}$. Applying the above formula recursively, we see that $\phi^*(t) = m_0(t/2) \prod_{k=1}^{\infty} m_0(t/2^k)\phi^*(0)$. This suggests that, given just the scaling coefficients, we can compute $\phi^*(t)$ and then take the inverse Fourier transform to find $\phi(t)$. An example of how to construct a father wavelet from a set of scaling coefficients is given in the next theorem.

9.18 Theorem. *Given coefficients* $\{\ell_k, \ k \in \mathbb{Z}\}$, *define a function*

$$m_0(t) = \frac{1}{\sqrt{2}} \sum_k \ell_k e^{-itk}. \tag{9.19}$$

Let

$$\phi^*(t) = \prod_{j=1}^{\infty} m_0\left(\frac{t}{2^j}\right) \tag{9.20}$$

and let ϕ *be the inverse Fourier transform of* ϕ^*. *Suppose that*

$$\frac{1}{\sqrt{2}} \sum_{k=N_0}^{N_1} \ell_k = 1 \tag{9.21}$$

for some $N_0 < N_1$, *that*

$$|m_0(t)|^2 + |m_0(t + \pi)|^2 = 1,$$

that $m_0(t) \neq 0$ *for* $|t| \leq \pi/2$, *and that there exists a bounded nonincreasing function* Φ *such that* $\int \Phi(|u|)du < \infty$ *and*

$$|\phi(x)| \leq \Phi(|x|)$$

for almost all x. *Then* ϕ *is a compactly supported father wavelet and* ϕ *is zero outside the interval* $[N_0, N_1]$.

Next, define W_k to be the orthogonal complement of V_k in V_{k+1}. In other words, each $f \in V_{k+1}$ can be written as a sum $f = v_k + w_k$ where $v_k \in V_k$, $w_k \in W_k$, and v_k and w_k are orthogonal. We write

$$V_{k+1} = V_k \bigoplus W_k.$$

Thus,

$$L_2(\mathbb{R}) = \overline{\bigcup_k V_k} = V_0 \bigoplus W_0 \bigoplus W_1 \bigoplus \cdots.$$

Define the **mother wavelet** by

$$\psi(x) = \sqrt{2} \sum_k (-1)^{k+1} \ell_{1-k}\, \phi(2x - k).$$

9.22 Theorem. *The functions $\{\psi_{jk}, k \in \mathbb{Z}\}$ form a basis for W_j. The functions*

$$\left\{ \phi_k, \psi_{jk},\ k \in \mathbb{Z}, j \in \mathbb{Z}_+ \right\} \tag{9.23}$$

are an orthonormal basis for $L_2(\mathbb{R})$. Hence, any $f \in L_2$ can be written as

$$f(x) = \sum_k \alpha_{0k}\, \phi_{0k}(x) + \sum_{j=0}^{\infty} \sum_k \beta_{jk}\, \psi_{jk}(x) \tag{9.24}$$

where

$$\alpha_{0k} = \int f(x)\phi_{0k}(x)dx \quad \text{and} \quad \beta_{jk} = \int f(x)\psi_{jk}(x)dx.$$

We have been denoting the first space in the MRA by V_0. This is simply a convention. We could just as well denote it by V_{j_0} for some integer j_0 in which case we write

$$f(x) = \sum_k \alpha_{j_0 k}\, \phi_{j_0 k}(x) + \sum_{j=j_0}^{\infty} \sum_k \beta_{jk}\, \psi_{jk}(x).$$

Of course, we still have not explained how to choose scaling coefficients to create a useful wavelet basis. We will not discuss the details here, but clever choices of scaling coefficients lead to wavelets with desirable properties. For example, in 1992 Ingrid Daubechies constructed a smooth, compactly supported "nearly" symmetric[1] wavelet called a **symmlet**. Actually, this is a

---

[1] There are no smooth, symmetric, compactly supported wavelets.

family of wavelets. A father symmlet of order N has support $[0, 2N-1]$ while the mother has support $[-N+1, N]$. The mother has N vanishing moments (starting with the 0th moment). The higher N is, the smoother the wavelet. There is no closed form expression for this wavelet (or for most wavelets) but it can be computed rapidly. Figure 9.5 shows the symmlet 8 mother wavelet which we will use in our examples. The scaling coefficients are:

```
 0.0018899503  -0.0003029205  -0.0149522583   0.0038087520
 0.0491371797  -0.0272190299  -0.0519458381   0.3644418948
 0.7771857517   0.4813596513  -0.0612733591  -0.1432942384
 0.0076074873   0.0316950878  -0.0005421323  -0.0033824160
```

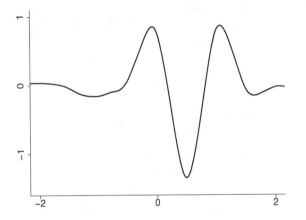

FIGURE 9.5. The symmlet 8 mother wavelet.

9.3 Wavelet Regression

Consider the regression problem

$$Y_i = r(x_i) + \sigma\epsilon_i \tag{9.25}$$

where $\epsilon_i \sim N(0,1)$ and $x_i = i/n$. We will further assume that $n = 2^J$ for some J. We will need to do some boundary corrections for x near 0 or 1; we defer that issue until later.

To estimate r using wavelets, we proceed as follows: First, approximate r with an expansion using n terms:

$$r(x) \approx r_n(x) = \sum_{k=0}^{2^{j_0}-1} \alpha_{j_0,k}\phi_{j_0,k}(x) + \sum_{j=j_0}^{J} \sum_{k=0}^{2^j-1} \beta_{jk}\psi_{jk}(x) \qquad (9.26)$$

where $\alpha_{j_0,k} = \int r(x)\phi_{j_0,k}(x)dx$ and $\beta_{jk} = \int r(x)\psi_{jk}(x)dx$. We refer to $\{\beta_{jk}, k = 0, 1, \ldots\}$ as the level-j coefficients. Form preliminary estimates of the coefficients:[2]

$$S_k = \frac{1}{n}\sum_i \phi_{j_0,k}(x_i)Y_i \quad \text{and} \quad D_{jk} = \frac{1}{n}\sum_i \psi_{jk}(x_i)Y_i \qquad (9.27)$$

which are called the **empirical scaling coefficients** and the **empirical detail coefficients**. Arguing as in the last chapter, we have that

$$S_k \approx N\left(\alpha_{j_0,k}, \frac{\sigma^2}{n}\right) \quad \text{and} \quad D_{jk} \approx N\left(\beta_{jk}, \frac{\sigma^2}{n}\right).$$

Next we use the following robust estimator for σ:

$$\widehat{\sigma} = \sqrt{n} \times \frac{\text{median}\left(|D_{J-1,k} - \text{median}(D_{J-1,k})| : \; k = 0, \ldots, 2^{J-1} - 1\right)}{0.6745}.$$

Inhomogeneous functions may have a few large wavelet coefficients even at the highest level J and this robust estimator should be relatively insensitive to such coefficients.

For the scaling coefficients, we take

$$\widehat{\alpha}_{j_0,k} = S_k.$$

---

[2]In practice, we do not compute S_k and D_{jk} using (9.27). Instead, we proceed as follows. The highest level scaling coefficients $\alpha_{J-1,k}$ are approximated by Y_k. This is reasonable since $\phi_{J-1,k}$ is highly localized and therefore

$$\mathbb{E}(Y_k) = f(k/n) \approx \int f(x)\phi_{J-1,k}(x)dx = \alpha_{J-1,k}.$$

Then we apply the cascade algorithm to get the rest of the coefficients; see the appendix for details. Some authors define

$$S_k = \frac{1}{\sqrt{n}}\sum_i \phi_{j_0,k}(x_i)Y_i \quad \text{and} \quad D_{jk} = \frac{1}{\sqrt{n}}\sum_i \psi_{jk}(x_i)Y_i$$

instead of using (9.27). This means that $S_k \approx N(\sqrt{n}\alpha_{j_0,k}, \sigma^2)$ and $D_{jk} \approx N(\sqrt{n}\beta_{jk}, \sigma^2)$. Hence, the final estimates should be divided by $\sqrt{n}$. Also, the estimate of the variance should be changed to

$$\widehat{\sigma} = \frac{\text{median}\left(|D_{J-1,k} - \text{median}(D_{J-1,k})| : \; k = 0, \ldots, 2^{J-1} - 1\right)}{0.6745}.$$

To estimate the coefficients β_{jk} of the mother wavelets we use a special type of shrinkage on the D_{jk} called **thresholding** which we describe in more detail in the next section. Finally, we plug the estimates into (9.26):

$$\widehat{r}_n(x) = \sum_{k=0}^{2^{j_0}-1} \widehat{\alpha}_{j_0,k}\phi_{j_0,k}(x) + \sum_{j=j_0}^{J} \sum_{k=0}^{2^j-1} \widehat{\beta}_{jk}\psi_{jk}(x).$$

9.4 Wavelet Thresholding

The wavelet regression method is the same as the procedure we used in the last chapter except for two changes: the basis is different and we use a different type of shrinkage called **thresholding** in which $\widehat{\beta}_{jk}$ is set to 0 if D_{jk} is small. Thresholding works better at finding jumps in the function than linear shrinkage. To see why, think of a function that is smooth except that it has some jumps in a few places. If we expand this function in a wavelet basis, the coefficients will be **sparse**, meaning that most will be small except for a few coefficients corresponding to the location of the jumps. Intuitively, this suggests setting most of the estimated coefficients to zero except the very large ones. This is precisely what a threshold rule does. More formally, we will see in the next section that thresholding shrinkage yields minimax estimators over large function spaces. Here are the details.

The estimates of the coefficients $\alpha_{j_0,k}$ of the father wavelet are equal to the empirical coefficients S_k; no shrinkage is applied. The estimates of the coefficients of the mother wavelets are based on shrinking the D_{jk}'s as follows. Recall that

$$D_{jk} \approx \beta_{jk} + \frac{\sigma}{\sqrt{n}}\epsilon_{jk}. \qquad (9.28)$$

The linear shrinkers we used in Chapters 7 and 8 were of the form $\widehat{\beta}_{jk} = cD_{jk}$ for some $0 \le c \le 1$. For wavelets, we use nonlinear shrinkage called **thresholding** which comes in two flavors: **hard thresholding** and **soft thresholding**. The hard threshold estimator is

$$\widehat{\beta}_{jk} = \left\{ \begin{array}{ll} 0 & \text{if } |D_{jk}| < \lambda \\ D_{jk} & \text{if } |D_{jk}| \ge \lambda. \end{array} \right. \qquad (9.29)$$

The soft threshold estimator is

$$\widehat{\beta}_{jk} = \left\{ \begin{array}{ll} D_{jk} + \lambda & \text{if } D_{jk} < -\lambda \\ 0 & \text{if } -\lambda \le D_{jk} < \lambda \\ D_{jk} - \lambda & \text{if } D_{jk} > \lambda \end{array} \right. \qquad (9.30)$$

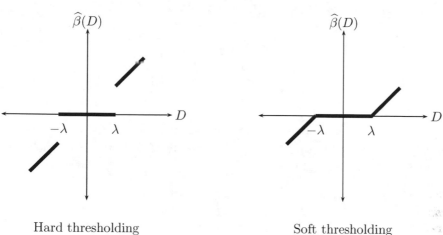

Hard thresholding   Soft thresholding

FIGURE 9.6. Hard and soft thresholding.

which can be written more succinctly as

$$\widehat{\beta}_{jk} = \text{sign}(D_{jk})(|D_{jk}| - \lambda)_+. \tag{9.31}$$

See Figure 9.6. In either case, the effect is to keep large coefficients and set the others equal to 0.

We will restrict attention to soft thresholding. We still need to choose the threshold λ. There are several methods for choosing λ. The simplest rule is the **universal threshold** defined by

$$\lambda = \widehat{\sigma}\sqrt{\frac{2\log n}{n}}. \tag{9.32}$$

To understand the intuition behind this rule, consider what happens when there is no signal, that is, when $\beta_{jk} = 0$ for all j and k. In this case, we would like all $\widehat{\beta}_{jk}$ to be 0 with high probability.

9.33 Theorem. *Suppose that $\beta_{jk} = 0$ for all j and k and let $\widehat{\beta}_{jk}$ be the soft-threshold estimator with the universal threshold (9.32). Then*

$$\mathbb{P}(\widehat{\beta}_{jk} = 0 \text{ for all } j, k) \to 1$$

as $n \to \infty$.

PROOF. To simplify the proof, assume that σ is known. Now, $D_{jk} \approx N(0, \sigma^2/n)$. Recall Mills' inequality: if $Z \sim N(0,1)$ then $\mathbb{P}(|Z| > t) \leq (c/t) e^{-t^2/2}$ where $c = \sqrt{2/\pi}$ is a constant. Thus,

$$\mathbb{P}(\max |D_{jk}| > \lambda) \leq \sum_{j,k} \mathbb{P}(|D_{jk}| > \lambda) \leq \sum_{j,k} \mathbb{P}\left(\frac{\sqrt{n}|D_{jk}|}{\sigma} > \frac{\sqrt{n}\lambda}{\sigma}\right)$$

$$= \sum_{j,k} \frac{c\sigma}{\lambda\sqrt{n}} \exp\left\{-\frac{1}{2}\frac{n\lambda^2}{\sigma^2}\right\} = \frac{c}{\sqrt{2\log n}} \to 0. \quad \blacksquare$$

More support for the thresholding rule is given by the following theorem which shows that soft thresholding does nearly as well as an "oracle" that chooses the best threshold.

9.34 Theorem (Donoho and Johnstone 1994). *Let*

$$Y_i = \theta_i + \frac{\sigma}{\sqrt{n}}\epsilon_i, \quad i = 1, \ldots, n$$

*where $\epsilon_i \sim N(0,1)$. For each $S \subset \{1, \ldots, n\}$ define the **kill it or keep it** estimator*

$$\widehat{\theta}_S = \left(X_1 I(1 \in S), \ldots, X_n I(n \in S)\right).$$

*Define the **oracle risk***

$$R_n^* = \min_S R(\widehat{\theta}_S, \theta) \tag{9.35}$$

where the minimum is over all kill it or keep it estimators, that is, S varies over all subsets of $\{1, \ldots, n\}$. Then

$$R_n^* = \sum_{i=1}^n \left(\theta_i^2 \wedge \frac{\sigma^2}{n}\right). \tag{9.36}$$

Further, if

$$\widehat{\theta} = (t(X_1), \ldots, t(X_n))$$

where $t(x) = \text{sign}(x)(|x| - \lambda_n)_+$ and $\lambda_n = \sigma\sqrt{2\log n/n}$, then for every $\theta \in \mathbb{R}^n$,

$$R_n^* \leq R(\widehat{\theta}, \theta) \leq (2\log n + 1)\left(\frac{\sigma^2}{n} + R_n^*\right). \tag{9.37}$$

Donoho and Johnstone call wavelet regression with the universal threshold, **VisuShrink**. Another estimator, called **SureShrink** is obtained by using a different threshold λ_j for each level. The threshold λ_j is chosen to minimize SURE (see Section 7.4) which, in this case, is

$$S(\lambda_j) = \sum_{k=1}^{n_j} \left[\frac{\widehat{\sigma}^2}{n} - 2\frac{\widehat{\sigma}^2}{n}I(|\widetilde{\beta}_{jk}| \leq \lambda_j) + \min(\widetilde{\beta}_{jk}^2, \lambda_j^2)\right] \tag{9.38}$$

where $n_j = 2^{j-1}$ is the number of parameters at level j. The minimization is performed over $0 \le \lambda_j \le (\hat{\sigma}/\sqrt{n_j})\sqrt{2 \log n_j}$.[3]

9.39 Example. The "blocks" function—introduced by Donoho and Johnstone (1995)—is defined by $r(x) = 3.655606 \times \sum_{j=1}^{11} h_j K(x - t_j)$ where

$$
\begin{aligned}
t &= (0.10, 0.13, 0.15, 0.23, 0.25, 0.40, 0.44, 0.65, 0.76, 0.78, 0.81) \\
h &= (4, -5, 3, -4, 5, -4.2, 2.1, 4.3, -3.1, 2.1, -4.2).
\end{aligned}
$$

The top left panel in Figure 9.7 shows $r(x)$. The top right plot shows 2048 data points generates from $Y_i = r(i/n) + \epsilon_i$ with $\epsilon \sim N(0, 1)$. The bottom left plot is the wavelet estimator using soft-thresholding with a universal threshold. We used a symmlet 8 wavelet. The bottom right plot shows a local linear fit with bandwidth chosen by cross-validation. The wavelet estimator is slightly better since the local linear fit has some extra wiggliness. However, the difference is not dramatic. There are situations where even this small difference might matter, such as in signal processing. But for ordinary nonparametric regression problems, there is not much practical difference between the estimators. Indeed, if we added confidence bands to these plots, they would surely be much wider than the difference between these estimators.

This is an easy example in the sense that the basic shape of the curve is evident from a plot of the data because the noise level is low. Let us now consider a noisier version of this example. We increase σ to 3 and we reduce the sample size to 256. The results are shown in Figure 9.8. It is hard to say which estimator is doing better here. Neither does particularly well. ∎

9.5 Besov Spaces

Wavelet threshold regression estimators have good optimality properties over **Besov spaces** which we now define. Let

$$
\Delta_h^{(r)} f(x) = \sum_{k=0}^{r} \binom{r}{k} (-1)^k f(x + kh).
$$

Thus, $\Delta_h^{(0)} f(x) = f(x)$ and

$$
\Delta_h^{(r)} f(x) = \Delta_h^{(r-1)} f(x + h) - \Delta_h^{(r-1)} f(x).
$$

---

[3]In practice, SureShrink is sometimes modified to include one more step. If the coefficients at level j are sparse, then universal thresholding is used instead. See Donoho and Johnstone (1995).

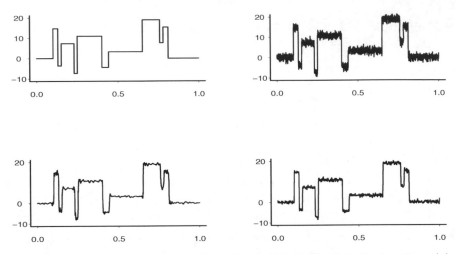

FIGURE 9.7. The "blocks" function from Example 9.39. Top left: the function $r(x)$. Top right: 2048 data points. Bottom left: $\widehat{r}_n$ using wavelets. Bottom right: $\widehat{r}_n$ using local linear regression with bandwidth chosen by cross-validation.

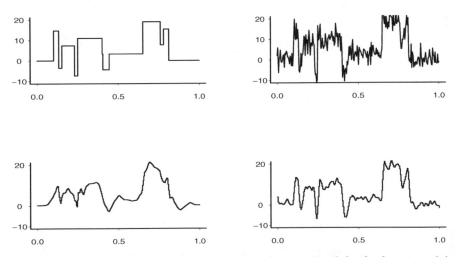

FIGURE 9.8. The "blocks" function from Example 9.39. Top left: the function $r(x)$. Top right: 256 data points. Bottom left: $\widehat{r}_n$ using wavelets. Bottom right: $\widehat{r}_n$ using local linear regression with bandwidth chosen by cross-validation.

Next define

$$w_{r,p}(f;t) = \sup_{|h| \leq t} ||\Delta_h^{(r)} f||_p$$

where $||g||_p = \left\{ \int |g(x)|^p dx \right\}^{1/p}$. Given (p,q,ς), let r be such that $r-1 \leq \varsigma \leq r$. The **Besov seminorm** is defined by

$$|f|_{p,q}^\varsigma = \left[\int_0^\infty (h^{-\varsigma} w_{r,p}(f;h))^q \frac{dh}{h} \right]^{1/q}.$$

For $q = \infty$ we define

$$|f|_{p,\infty}^\varsigma = \sup_{0 < h < 1} \frac{w_{r,p}(f;h)}{h^\varsigma}.$$

The **Besov space** $B_{p,q}^\varsigma(c)$ is defined to be the set of functions f mapping $[0,1]$ into $\mathbb{R}$ such that $\int |f|^p < \infty$ and $|f|_{p,q}^\varsigma \leq c$.

This definition is hard to understand but here are some special cases. The Sobolev space $W(m)$ (see Definition 7.8) corresponds to the Besov ball $B_{2,2}^m$. The generalized Sobolev space $W_p(m)$ which uses an L_p norm on the mth derivative is almost a Besov space in the sense that $B_{p,1}^m \subset W_p(m) \subset B_{p,\infty}^m$. Let $s = m + \delta$ for some integer m and some $\delta \in (0,1)$. The **Hölder space** is the set of bounded functions with bounded m^{th} derivative such that $|f^m(u) - f^m(t)| \leq |u - t|^\delta$ for all u, t. This space is equivalent to $B_{\infty,\infty}^{m+\delta}$. The set T consisting of functions of bounded variation satisfies: $B_{1,1}^1 \subset T \subset B_{1,\infty}^1$. Thus, Besov spaces include a wide range of familiar function spaces.

It is easier to understand Besov spaces in terms of the coefficients of the wavelet expansion. If the wavelets are sufficiently smooth then the wavelet coefficients β of a function $f \in B_{p,q}^\varsigma(c)$ satisfy $||\beta||_{p,q}^\varsigma \leq c$, where

$$||\beta||_{p,q}^\varsigma = \left(\sum_{j=j_0}^\infty \left(2^{j(\varsigma+(1/2)-(1/p))} \left(\sum_k |\beta_{jk}|^p \right)^{1/p} \right)^q \right)^{1/q}. \tag{9.40}$$

In the theorem that follows, we use the following notation: write $a_n \asymp b_n$ to mean that a_n and b_n go to 0 at the same rate. Formally,

$$0 < \liminf_{n \to \infty} \left| \frac{a_n}{b_n} \right| \leq \limsup_{n \to \infty} \left| \frac{a_n}{b_n} \right| < \infty.$$

9.41 Theorem (Donoho–Johnstone 1995). *Let $\hat{r}_n$ be the SureShrink estimator. Let ψ have r null moments and r continuous derivatives where $r > \max\{1, \varsigma\}$. Let $R_n(p,q,\varsigma,C)$ denote the minimax risk over the Besov ball $B_{p,q}^\varsigma(c)$. Then*

$$\sup_{r \in B_{p,q}^\varsigma(C)} \frac{1}{n} \mathbb{E} \left(\sum_i (\hat{r}_n(x_i) - r(x_i))^2 \right) \asymp R_n(p,q,\varsigma,c)$$

for all $1 \le p, q \le \infty$, $C \in (0, \infty)$ and $\varsigma_0 < \varsigma < r$ where

$$\varsigma_0 = \max \left\{ \frac{1}{p}, 2 \left(\frac{1}{p} - \frac{1}{2} \right)^+ \right\}.$$

No linear estimator achieves the optimal rate over this range of spaces. The universal shrinkage rule also attains the minimax rate, up to factors involving $\log n$.

The theorem says that wavelet estimators based on threshold rules achieve the minimax rate, simultaneously over a large set of Besov spaces. Other estimators, such as local regression estimators with a constant bandwidth, do not have this property.

9.6   Confidence Sets

At the time of this writing, practical, simultaneous confidence bands for wavelet estimators are not available. An asymptotic pointwise method is given in Picard and Tribouley (2000). Confidence balls are available from Genovese and Wasserman (2005).

For a Besov space $B_{p,q}^\varsigma$, let

$$\gamma = \left\{ \begin{array}{ll} \varsigma & p \ge 2 \\ \varsigma + \frac{1}{2} - \frac{1}{p} & 1 \le p < 2. \end{array} \right. \tag{9.42}$$

Let $B(p, q, \gamma)$ be the set of wavelet coefficients corresponding to functions in the Besov space. We assume that the mother and father wavelets are bounded, have compact support, and have derivatives with finite L_2 norms. Let $\mu^n = (\mu_1, \ldots, \mu_n)$ be the first n wavelet coefficients strung out as a single vector.

9.43 Theorem (Genovese and Wasserman 2004). *Let $\widehat{\mu}$ be the estimated wavelet coefficients using the universal soft threshold $\lambda = \widehat{\sigma}\sqrt{\log n / n}$. Define*

$$D_n = \left\{ \mu^n : \sum_{\ell=1}^n (\mu_\ell - \widehat{\mu}_\ell)^2 dx \le s_n^2 \right\} \tag{9.44}$$

where

$$s_n^2 = \sqrt{2}\sigma^2 \frac{z_\alpha}{\sqrt{n}} + S_n(\lambda), \tag{9.45}$$

$$S_n(\lambda) = \frac{\widehat{\sigma}^2}{n} 2^{j_0} + \sum_{j=j_0}^J S(\lambda_j) \tag{9.46}$$

and

$$S_j(\lambda_j) = \sum_{k=1}^{n_j} \left[\frac{\sigma^2}{n} - 2\frac{\sigma^2}{n} I(|\widetilde{\beta}_{jk}| \leq \lambda_j) + \min(\widetilde{\beta}_{jk}^2, \lambda_j^2) \right].$$

Then, for any $\delta > 0$,

$$\lim_{n\to\infty} \sup_{\mu\in\Delta(\delta)} |\mathbb{P}(\mu^n \in D_n) - (1-\alpha)| = 0 \tag{9.47}$$

where

$$\Delta(\delta) = \bigcup \left\{ B(p,q,\gamma) : \ p \geq 1, \ q \geq 1, \ \gamma > (1/2) + \delta \right\}.$$

9.7 Boundary Corrections and Unequally Spaced Data

If the data live on an interval but the wavelet basis lives on $\mathbb{R}$, then a correction is needed since the wavelets will not usually be orthonormal when restricted to the interval. The simplest approach is to **mirror** the data. The data are repeated in reverse order around the endpoint. Then the previously discussed methods are applied.

When the data are not equally spaced or n is not a power of 2, we can put the data into equally spaced bins and average the data over the bins. We choose the bins as small as possible subject to having data in each bin and subject to the number of bins m being of the form $m = 2^k$ for some integer k.

For other approaches to these problems see Härdle et al. (1998) as well as the references therein.

9.8 Overcomplete Dictionaries

Although wavelet bases are very flexible, there are cases where one might want to build even richer bases. For example, one might consider combining several bases together. This leads to the idea of a **dictionary**.

Let $Y = (Y_1, \ldots, Y_n)^T$ be a vector of observations where $Y_i = r(x_i) + \epsilon_i$. Let D be an $n \times m$ matrix with $m > n$. Consider estimating r using $D\beta$ where $\beta = (\beta_1, \ldots, \beta_m)^T$. If m were equal to n and the columns of D were orthonormal, we are back to orthogonal basis regression as in this and the last chapter. But when $m > n$, the columns can no longer be orthogonal and we say that the dictionary is **overcomplete.** For example, we might take D to have $m = 2n$ columns: the first n columns being the basis elements of a cosine basis and the second n columns being the basis elements of a "spike" basis. This would enable us to estimate a function that is "smooth plus spikes."

There is growing evidence—theoretical and practical—that the method called **basis pursuit**, due to Chen et al. (1998), leads to good estimators. In this method, one chooses $\widehat{\beta}$ to minimize

$$||Y - D\beta||_2^2 + \lambda||\beta||_1$$

where $|| \cdot ||_2$ denotes the L_2 norm, $|| \cdot ||_1$ denotes the L_1 norm and $\lambda > 0$ is a constant. Basis pursuit is related to the regression variable selection techniques called **the lasso** (Tibshirani (1996)) and **LARS** (Efron et al. (2004)).

9.9   Other Adaptive Methods

There are many other spatially adaptive methods besides wavelets; these include variable bandwidth kernel methods (Lepski et al. (1997), Müller and Stadtmller (1987)), local polynomials (Loader (1999a) and Fan and Gijbels (1996)) variable knot splines (Mammen and van de Geer (1997)) among others. Here we outline an especially simple, elegant method due to Goldenshluger and Nemirovski (1997) called intersecting confidence intervals (ICI).

Consider estimating a regression function $r(x)$ at a point x and let $Y_i = r(x_i) + \epsilon_i$ where $\epsilon_i \sim N(0, \sigma^2)$. To simplify the discussion, we will assume that σ is known. The idea is to construct a confidence interval for $r(x)$ using a sequence of increasing bandwidths h. We will then choose the first bandwidth at which the intervals do not intersect. See Figure 9.9.

Let

$$\widehat{r}_h(x) = \sum_{i=1}^{n} Y_i \ell_i(x, h)$$

be a linear estimator depending on bandwidth h. Let the bandwidth h vary in a finite set $h_1 < \cdots < h_n$ and let $\widehat{r}_j = \widehat{r}_{h_j}(x)$. For example, with equally spaced data on $[0, 1]$ we might take $h_j = j/n$. We can write

$$\widehat{r}_j = \sum_{i=1}^{n} r(x_i)\ell_i(x, h_j) + \xi_j$$

where

$$\xi_j = \sum_{i=1}^{n} \epsilon_i \ell_i(x, h_j) \sim N(0, s_j^2), \quad s_j = \sigma\sqrt{\sum_{i=1}^{n} \ell_i^2(x, h_j)}.$$

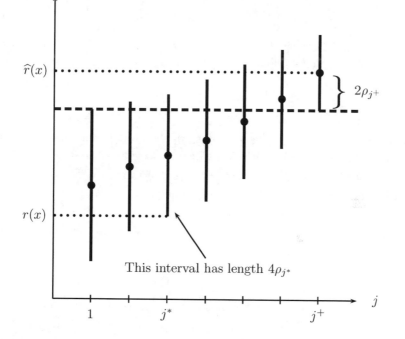

FIGURE 9.9. The Goldenshluger and Nemirovski ICI method. All the intervals $\{D_j : j \leq j^*\}$ contain the true value $r(x)$. All the intervals $\{D_j :\ j \leq j^+\}$ overlap. The estimate $\widehat{r}(x)$ is no more than $2\rho_{j^+} + 4\rho_{j^*} \leq 6\rho_{j^*}$ away from $r(x)$.

Let $\rho_j = \kappa s_j$ where $\kappa > \sqrt{2 \log n}$. Then

$$\mathbb{P}\big(\max_j |\xi_j| > \rho_j,\ \text{ for some } j\big) \leq nP\big(|N(0,1)| > \kappa\big) \to 0 \text{ as } n \to \infty.$$

So $|\xi_j| \leq \rho_j$ for all j expect on a set of probability tending to 0. For the rest of the argument, assume that $|\xi_j| \leq \rho_j$ for all j.

Form n intervals

$$D_j = \big[\widehat{r}_j(x) - 2\rho_j,\ \widehat{r}_j(x) + 2\rho_j\big], \quad j = 1, \ldots, n.$$

The adaptive estimator is defined to be

$$\widehat{r}(x) = \widehat{r}_{j^+}(x) \tag{9.48}$$

where j_+ is the largest integer such that the intervals overlap:

$$j_+ = \max\left\{ k : \bigcap_{j=1}^{k} D_j \neq \emptyset \right\}. \tag{9.49}$$

We now outline why $\widehat{r}(x)$ is adaptive.

Let $\overline{r}_j = \mathbb{E}(\widehat{r}_j)$ and note that

$$|\widehat{r}_j - r(x)| \leq |\widehat{r}_j - \overline{r}_j| + |\overline{r}_j - r(x)| = b_j + |\xi_j|$$

where $b_j = |\overline{r}_j - \overline{r}_j|$ is the bias. Assume—as is the case for most smoothers—that b_j decreases in j while $s_j^2 = \mathbb{V}(\xi_j)$ increases in j. Let

$$j^* = \max\{j : \ b_j \leq \rho_j\}.$$

The bandwidth h_{j^*} balances the bias and variance. The risk of the estimator with bandwidth h_{j^*} is of order ρ_{j^*}. Thus, ρ_{j^*} is the risk of an oracle estimator that knows the best bandwidth. We will refer to ρ_{j^*} as the oracle risk.

For $j \leq j^*$, $|\widehat{r}_j - r(x)| \leq b_j + \rho_j \leq 2\rho_j$. Hence, $r(x) \in D_j$ for all $j \leq j^*$. In particular, all D_j, $j \leq j_*$ have at least one point in common, namely, $r(x)$. By the definition of j^+ it follows that $j^* \leq j^+$. Also, every point in D_{j^*} is at most $4\rho_{j^*}$ away from $r(x)$. Finally, $D_{j^+} \cap D_{j^*} \neq \emptyset$ (from the definition of j^*) and D_{j^+} has half-length $2\rho_{j^+}$. Thus,

$$|\widehat{r}(x) - r(x)| \leq |\widehat{r}(x) - \widehat{r}_{j^*}| + |\widehat{r}_{j^*} - r(x)| \leq 2\rho_{j^+} + 4\rho_{j^*} \leq 6\rho_{j^*}.$$

We conclude that, with probability tending to 1, $|\widehat{r}(x) - r(x)| \leq 6\rho_{j^*}$.

A specific implementation of the idea is as follows. Fit a polynomial of order m over an interval Δ containing the point $x \in (0,1)$. The resulting estimator is

$$\widehat{r}_\Delta(x) = \sum_i \alpha_\Delta(x_i, x) Y_i$$

for weights $\alpha_\Delta(x_i, x)$. The weights can be written as

$$\alpha_\Delta(x, u) = \sum_{j=0}^m q_\Delta^{(j)}(x) \left(\frac{u - a}{b - a}\right)^j$$

where $a = \min\{x_i : \ x_i \in \Delta\}$ and $b = \max\{x_i : \ x_i \in \Delta\}$. It can be shown that

$$|\alpha_\Delta(x_i)| \leq \frac{c_m}{N_\Delta}$$

where N_δ is the number of points in Δ and c_m is a constant depending only on m. Also, it can be shown that the quantity

$$\tau_m \equiv \frac{N_\Delta}{\theta_m} \max_{i,j} |q_\Delta^{(j)}(x_i)|$$

depends only on m. Let

$$D_\Delta = \left[\widehat{r}_\Delta - 2\kappa s_\Delta, \ \widehat{r}_\Delta + 2\kappa s_\Delta\right]$$

where

$$s_\Delta = \left(\sum_i \alpha_\Delta^2(x_i) \right)^{1/2}$$

and

$$\kappa_n = 2\sigma\sqrt{(m+2)\log n}\, 2^m(m+1)c_m^2\tau_m.$$

Now let $\widehat{\Delta}$ be the largest interval Δ containing x such that

$$\bigcap_{\substack{\delta \in \mathcal{D} \\ \delta \subset \Delta}} D_\delta \neq \emptyset$$

where $\mathcal{D}$ denotes all intervals containing x. Finally let $\widehat{r}(x) = \widehat{r}_{\widehat{\Delta}}(x)$.

Now we suppose that r is ℓ times differentiable on an interval $\Delta_0 = [x - \delta_0, x + \delta_0] \subset [0,1]$ for some $\ell \geq 1$. Also, assume that

$$\left(\int_{\Delta_0} |r^{(\ell)}(t)|^p dt \right)^{1/p} \leq L$$

for some $L > 0$, and $p \geq 1$, where

$$\ell \leq m + 1, \quad p\ell > 1.$$

Then we have[4] the following:

9.50 Theorem (Goldenshluger and Nemirovski 1997). *Under the conditions above,*

$$\left(\mathbb{E}|\widehat{r}(x) - r(x)|^2 \right)^{1/2} \leq CB_n \tag{9.51}$$

for some constant $C > 0$ that only depends on m, where

$$B_n = \left[\left(\frac{\log n}{n} \right)^{(p\ell-1)/(2p\ell+p-2)} L^{p/(2p\ell+p-2)} + \sqrt{\frac{\log n}{n\delta_0}} \right]. \tag{9.52}$$

The right-hand side of (9.51) is, except for logarithmic factors, the best possible risk. From the results of Lepskii (1991), the logarithmic factor is unavoidable. Since the estimator did not make use of the smoothness parameters p, ℓ and L, this means that the estimator adapts to the unknown smoothness.

Let us also briefly describe (a version of) the method of Lepski et al. (1997). Let $\mathcal{H} = \{h_0, h_1, \ldots, h_m\}$ be a set of bandwidths, where $h_j = a^{-j}$ where $a > 1$

---

[4]I've stated the result in a very specific form. The original result is more general than this.

(they use $a = 1.02$) and m is such that $h_m \approx \sigma^2/n$. Let $\widehat{r}_h(x)$ be the kernel estimator based on bandwidth h and kernel K. For each bandwidth h, we test the hypothesis that further reducing h does not significantly improve the fit. We take $\widehat{h}$ to be the largest bandwidth for which this test does not reject. Specifically,

$$\widehat{h} = \max\left\{ h \in \mathcal{H} : \; |\widehat{r}_h(x) - \widehat{r}_\eta(x)| \leq \psi(h, \eta), \;\; \text{for all } \eta < h, \eta \in \mathcal{H}\right\}$$

where

$$\psi(h, \eta) = \frac{D\sigma}{n\eta}\sqrt{1 + \log\left(\frac{1}{\eta}\right)}$$

and $D > 2(1 + ||K||\sqrt{14})$. They show that this bandwidth selection method yields to an estimator that adapts over a wide range of Besov spaces.

9.10   Do Adaptive Methods Work?

Let us take a closer look at the idea of spatial adaptation following some ideas in Donoho and Johnstone (1994). Let $A_1, \ldots, A_L$ be intervals that partition $[0, 1]$:

$$A_1 = [a_0, a_1), \;\; A_2 = [a_1, a_2), \;\; \ldots, \;\; A_L = [a_{L-1}, a_L]$$

where $a_0 = 0$ and $a_L = 1$. Suppose that r is a piecewise polynomial, so that $r(x) = \sum_{\ell=1}^{L} p_\ell(x) I(x \in A_\ell)$ where p_ℓ is a polynomial of degree D on the set A_ℓ. If the breakpoints $a = (a_1, \ldots, a_L)$ and the degree D are known, then we can fit a polynomial of degree D over each A_ℓ using least squares. This is a parametric problem and the risk is of order $O(1/n)$. If we do not know the breakpoints a and we fit a kernel regression then it can be shown that the risk is in general not better than order $O(1/\sqrt{n})$ due to the possibility of discontinuities at the breakpoints. In contrast, Donoho and Johnstone show that the wavelet method has risk of order $O(\log n/n)$. This is an impressive theoretical achievement.

On the other hand, examples like those seen here suggest that the practical advantages are often quite modest. And from the point of view of inference (which features in the estimate are real?) the results in Chapter 7 show that wavelet confidence balls cannot converge faster than $n^{-1/4}$. To see this, consider the piecewise polynomial example again. Even knowing that r is a piecewise polynomial, it still follows that the vector $(r(x_1), \ldots, r(x_n))$ can take any value in $\mathbb{R}^n$. It then follows from Theorem 7.71 that no confidence ball can shrink faster than $n^{-1/4}$. Thus, we are in the peculiar situation that the function estimate may converge quickly but the confidence set converges slowly.

So, do adaptive methods work or not? If one needs accurate function estimates and the noise level σ is low, then the answer is that adaptive function estimators are very effective. But if we are facing a standard nonparametric regression problem and we are interested in confidence sets, then adaptive methods do not perform significantly better than other methods such as fixed bandwidth local regression.

9.11 Bibliographic Remarks

A good introduction to wavelets is Ogden (1997). A more advanced treatment can be found in Härdle et al. (1998). The theory of statistical estimation using wavelets has been developed by many authors, especially David Donoho and Iain Johnstone. The main ideas are in the following remarkable series of papers: Donoho and Johnstone (1994), Donoho and Johnstone (1995), Donoho et. al. (1995), and Donoho and Johnstone (1998). The material on confidence sets is from Genovese and Wasserman (2005).

9.12 Appendix

Localization of Wavelets. The idea that wavelets are more localized than sines and cosines can be made precise. Given a function f, define its radius to be

$$\Delta_f = \frac{1}{||f||} \left\{ \int (x - \overline{x})^2 |f(x)|^2 dx \right\}^{1/2}$$

where

$$\overline{x} = \frac{1}{||f||^2} \int x |f(x)|^2 dx.$$

Imagine drawing a rectangle on the plane with sides of length Δ_f and Δ_{f^*}. For a function like a cosine, this is a rectangle with 0 width on the y-axis and infinite width in the x-axis. We say that a cosine is localized in frequency but nonlocalized in space. In contrast, a wavelet has a rectangle with finite length in both dimensions. Hence, wavelets are localized in frequency and space. There is a limit to how well we can simultaneously localize in frequency and space.

9.53 Theorem (Heisenberg's uncertainty relation). *We have that*

$$\Delta_f \Delta_{f^*} \geq \frac{1}{2} \tag{9.54}$$

with equality when f is a Normal density.

This inequality is called the **Heisenberg uncertainty principle** since it first appeared in the physics literature when Heisenberg was developing quantum mechanics.

Fast Computations for Wavelets. The scaling coefficients make it easy to compute the wavelet coefficients. Recall that

$$\alpha_{jk} = \int f(x)\phi_{jk}(x)dx \ \text{ and } \ \beta_{jk} = \int f(x)\psi_{jk}(x)dx.$$

By definition, $\phi_{jk}(x) = 2^{j/2}\phi(2^j x - k)$ and by (9.16), $\phi(2^j x - k) = \sum_r \ell_r \phi_{1,r}(2^j x - k)$. Hence,

$$\begin{aligned}
\phi_{jk}(x) &= \sum_r \ell_r 2^{j/2}\phi_{1,r}(2^j x - k) = \sum_r \ell_r 2^{(j+1)/2}\phi(2^{j+1}x - 2k - r) \\
&= \sum_r \ell_r \phi_{j+1,\ell+2k}(x) = \sum_r \ell_{r-2k}\phi_{j+1,r}(x).
\end{aligned}$$

Thus,

$$\begin{aligned}
\alpha_{jk} &= \int f(x)\phi_{jk}(x)dx = \int f(x)\sum_r \ell_{r-2k}\phi_{j+1,r}(x)dx \\
&= \sum_r \ell_{r-2k} \int f(x)\phi_{j+1,r}(x)dx = \sum_r \ell_{r-2k}\,\alpha_{j+1,r}.
\end{aligned}$$

By similar calculations, one can show that $\beta_{jk} = \sum_r (-1)^{r+1}\ell_{-r+2k+1}\,\alpha_{j+1,r}$. In summary we get the following **cascade equations**:

The Cascade Equations

$$\alpha_{jk} = \sum_r \ell_{r-2k}\,\alpha_{j+1,r} \tag{9.55}$$

$$\beta_{jk} = \sum_r (-1)^{r+1}\ell_{-r+2k+1}\,\alpha_{j+1,r}. \tag{9.56}$$

Once we have the scaling coefficients $\{\alpha_{jk}\}$ for some J, we may determine $\{\alpha_{jk}\}$ and $\{\beta_{jk}\}$ for all $j < J$ by way of the cascade equations (9.55) and

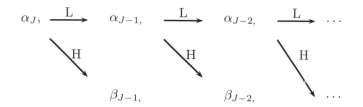

FIGURE 9.10. The cascade algorithm.

(9.56). This method of computing the coefficients is called the **pyramid algorithm** or the **cascade algorithm**.

In regression problems, we will use the data $Y_1, \ldots, Y_n$ to approximate the scaling coefficients at some high-level J. The other coefficients are found from the pyramid algorithm. This process is called the **discrete wavelet transform**. It only requires $O(n)$ operations for n data points.

These ideas can be expressed in the language of signals and filters. A **signal** is defined to be a sequence $\{f_k\}_{k \in \mathbb{Z}}$ such that $\sum_k f_k^2 < \infty$. A **filter** A is a function that operates on signals. A filter A can be represented by some coefficients $\{a_k\}_{k \in \mathbb{Z}}$ and the action of A on a signal f—called **discrete convolution**—produces a new signal denoted by Af whose k^{th} coefficient is

$$(Af)_k = \sum_r a_{r-2k} f_r. \tag{9.57}$$

Let $\alpha_j = \{\alpha_{jk}\}_{k \in \mathbb{Z}}$ be the scaling coefficients at level j. Let L be the filter with coefficients $\{\ell_k\}$. L is called a **low-pass filter**. From Equation (9.55) we get

$$\alpha_{j-1} = L\alpha_j, \quad \text{and} \quad \alpha_{j-m} = L^m \alpha_j, \tag{9.58}$$

where L^m means: apply the filter m times. Let H be the filter with coefficients $h_k = (-1)^{k+1} \ell_{1-k}$. Then (9.56) implies that

$$\beta_{j-1} = H\alpha_j, \quad \text{and} \quad \beta_{j-m} = HL^{m-1}\alpha_j. \tag{9.59}$$

H is called a **high-pass** filter. See Figure 9.10 for a schematic representation of the algorithm.

9.13 Exercises

1. Show that, for the Haar wavelet, $\ell_0 = \ell_1 = 2^{-1/2}$ and $\ell_k = 0$ for $k \neq 0, 1$.

2. Prove Theorem 9.6.

3. Prove that the Haar wavelets form an MRA; see Example 9.15.

4. Prove equation (9.17).

5. Generate data

$$Y_i = r(i/n) + \sigma \epsilon_i$$

where r is the doppler function, $n = 1024$, and $\epsilon_i \sim N(0, 1)$.

(a) Use wavelets to fit the curve. Try the following shrinkage methods: (i) James–Stein applied to each resolution level, (ii) universal shrinkage and (iii) SureShrink. Try $\sigma = .01$, $\sigma = .1$, $\sigma = 1$. Compare the function estimates to the REACT method and to local linear regression.

(b) Repeat (a) but add outliers to the ϵ_i by generating them from:

$$\epsilon_i \sim .95 \ N(0, 1) + .05 \ N(0, 4).$$

How does this affect your results?

6. Let $X_1, \ldots, X_n \sim f$ for some density f on $[0, 1]$. Let's consider constructing a wavelet histogram. Let ϕ and ψ be the Haar father and mother wavelet. Write

$$f(x) \approx \phi(x) + \sum_{j=0}^{J} \sum_{k=0}^{2^j - 1} \beta_{jk} \psi_{jk}(x)$$

where $J \approx \log_2(n+1)$. The total number of wavelet coefficients is about n. Now,

$$\beta_{jk} = \int_0^1 \psi_{jk}(x) f(x) dx = E_f \left[\psi_{jk}(X) \right].$$

An unbiased estimator of β_{jk} is

$$\widetilde{\beta}_{jk} = \frac{1}{n} \sum_{i=1}^{n} \psi_{jk}(X_i).$$

(a) For $x < y$ define

$$N_{x,y} = \sum_{i=1}^{n} I(x \leq X_i < y).$$

Show that

$$\widetilde{\beta}_{jk} = \frac{2^{j/2}}{n} \left(N_{2k/(2^{j+1}),(2k+1)/(2^{j+1})} - N_{(2k+1)/(2^{j+1}),(2k+2)/(2^{j+1})} \right).$$

(b) Show that
$$\widetilde{\beta}_{jk} \approx N(\beta_{jk}, \sigma_{jk}^2).$$
Find an expression for σ_{jk}.

(c) Consider the shrinkage estimator $\widehat{\beta}_{jk} = a_j \widetilde{\beta}_{jk}$. (Same shrinkage coefficient across each level of wavelets.) Take σ_{jk} as known, and find a_j to minimize Stein's unbiased risk estimate.

(d) Now find an estimate of σ_{jk}. Insert this estimate into your formula for a_j. You now have a method for estimating the density. Your final estimate is
$$\widehat{f}(x) = \phi(x) + \sum_j \sum_k \widehat{\beta}_{jk} \psi_{jk}(x).$$
Try this on the geyser duration data from the book website. Compare it to the histogram with binwidth chosen using cross-validation.

(e) Notice that $\widetilde{\beta}_{jk}$ is $2^{j/2}$ times the difference of two sample proportions. Furthermore, $\beta_{jk} = 0$ corresponds to the two underlying Binomial parameters being equal. Hence, we can form a hard threshold estimator by testing (at some level α) whether each $\beta_{jk} = 0$ and only keeping those that are rejected. Try this for the data in part (d). Try different values of α.

7. Let R_n^* be the oracle risk defined in equation (9.35).

   (a) Show that
   $$R_n^* = \sum_{i=1}^n \left(\frac{\sigma^2}{n} \wedge \theta_i^2 \right)$$
   where $a \wedge b = \min\{a, b\}$.

   (b) Compare R_n^* to the Pinsker bound in Theorem 7.28. For which vectors $\theta \in \mathbb{R}^n$ is R_n^* smaller than than the Pinsker bound?

8. Find an exact expression for the risk of the hard threshold rule and the soft threshold rule (treating σ as known).

9. Generate data
   $$Y_i = r(i/n) + \sigma \epsilon_i$$
   where r is the Doppler function, $n = 1024$, and $\epsilon_i \sim N(0, 1)$ and $\sigma = 0.1$. Apply the ICI method from Section 9.9 to estimate r. Use a kernel estimator and take the grid of bandwidths to be $\{1/n, \ldots, 1\}$. First, take σ as known. Then estimate σ using one of the methods from Section 5.6. Now apply the method of Lepski et al. (1997) also from Section 5.6.

10
Other Topics

In this chapter we mention some other issues related to nonparametric inference including: measurement error, inverse problems, nonparametric Bayesian inference, semiparametric inference, correlated errors, classification, sieves, shape-restricted inference, testing, and computation.

10.1 Measurement Error

Suppose we are interested in regressing the outcome Y on a covariate X but we cannot observe X directly. Rather, we observe X plus noise U. The observed data are $(X_1^\bullet, Y_1), \ldots, (X_n^\bullet, Y_n)$ where

$$
\begin{aligned}
Y_i &= r(X_i) + \epsilon_i \\
X_i^\bullet &= X_i + U_i, \qquad \mathbb{E}(U_i) = 0.
\end{aligned}
$$

This is called a **measurement error** problem or an **errors-in-variables** problem. A good reference is Carroll et al. (1995) which we follow closely in this section.

The model is illustrated by the directed graph in Figure 10.1. It is tempting to ignore the error and just regress Y on $X^\bullet$ but this leads to inconsistent estimates of $r(x)$.

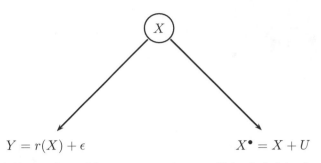

FIGURE 10.1. Regression with measurement error. X is circled to show that it is not observed. $X^\bullet$ is a noisy version of X. If you regress Y on $X^\bullet$, you will get an inconsistent estimate of $r(x)$.

Before discussing the nonparametric problem, we first consider the linear regression version of this problem. The model is

$$
\begin{aligned}
Y_i &= \beta_0 + \beta_1 X_i + \epsilon_i \\
X_i^\bullet &= X_i + U_i.
\end{aligned}
$$

Let $\sigma_x^2 = \mathbb{V}(X)$, and assume that ϵ is independent of X, has mean 0 and variance σ_ϵ^2. Also assume that U is independent of X, with mean 0 and variance σ_u^2. Let $\widehat{\beta}_1$ be the least squares estimator of β_1 obtained by regressing the Y_i's on the $X_i^\bullet$'s. It can be shown (see Exercise 2) that

$$
\widehat{\beta} \xrightarrow{\text{a.s.}} \lambda \beta_1 \tag{10.1}
$$

where

$$
\lambda = \frac{\sigma_x^2}{\sigma_x^2 + \sigma_u^2} < 1. \tag{10.2}
$$

Thus, the effect of the measurement is to bias the estimated slope towards 0, an effect that is usually called **attenuation bias**. Staudenmayer and Ruppert (2004) showed that a similar result holds for nonparametric regression. If we use local polynomial regression (with odd degree polynomial) and do not account for the measurement error, the estimator $\widehat{r}_n(x)$ has, asymptotically, an excess bias of

$$
\sigma_u^2 \left(\frac{f'(x)}{f(x)} r'(x) + \frac{r''(x)}{2} \right) \tag{10.3}
$$

where f is the density of X.

Returning to the linear case, if there are several observed values of $X^\bullet$ for each X then σ_u^2 can be estimated. Otherwise, σ_u^2 must be estimated by external means such as through background knowledge of the noise mechanism. For our purposes, we will assume that σ_u^2 is known. Since, $\sigma_\bullet^2 = \sigma_x^2 + \sigma_u^2$, we can estimate σ_x^2 by

$$\widehat{\sigma}_x^2 = \widehat{\sigma}_\bullet^2 - \sigma_u^2 \tag{10.4}$$

where $\widehat{\sigma}_\bullet^2$ is the sample variance of the $X_i^\bullet$s. Plugging these estimates into (10.2), we get an estimate $\widehat{\lambda} = (\widehat{\sigma}_\bullet^2 - \sigma_u^2)/\widehat{\sigma}_\bullet^2$ of λ. An estimate of β_1 is

$$\widetilde{\beta}_1 = \frac{\widehat{\beta}_1}{\widehat{\lambda}} = \frac{\widehat{\sigma}_\bullet^2}{\widehat{\sigma}_\bullet^2 - \sigma_u^2}\widehat{\beta}_1. \tag{10.5}$$

This is called the **method of moments estimator**. See Fuller (1987) for more details.

Another method for correcting the attenuation bias is SIMEX which stands for **simulation extrapolation** and is due to Cook and Stefanski (1994) and Stefanski and Cook (1995). Recall that the least squares estimate $\widehat{\beta}_1$ is a consistent estimate of

$$\frac{\beta_1\sigma_x^2}{\sigma_x^2 + \sigma_u^2}.$$

Generate new random variables

$$\widetilde{X}_i = X_i^\bullet + \sqrt{\rho}\,U_i$$

where $U_i \sim N(0,1)$. The least squares estimate obtained by regressing the Y_i's on the $\widetilde{X}_i$'s is a consistent estimate of

$$\Omega(\rho) = \frac{\beta_1\sigma_x^2}{\sigma_x^2 + (1+\rho)\sigma_u^2}. \tag{10.6}$$

Repeat this process B times (where B is large) and denote the resulting estimators by $\widehat{\beta}_{1,1}(\rho), \ldots, \widehat{\beta}_{1,B}(\rho)$. Then define

$$\widehat{\Omega}(\rho) = \frac{1}{B}\sum_{b=1}^{B}\widehat{\beta}_{1,b}(\rho).$$

Now comes some clever sleight of hand. Setting $\rho = -1$ in (10.6) we see that $\Omega(-1) = \beta_1$ which is the quantity we want to estimate. The idea is to compute $\widehat{\Omega}(\rho)$ for a range of values of ρ such as $0, 0.5, 1.0, 1.5, 2.0$. We then extrapolate the curve $\widehat{\Omega}(\rho)$ back to $\rho = -1$; see Figure 10.2. To do the extrapolation, we fit the values $\widehat{\Omega}(\rho_j)$ to the curve

$$G(\rho; \gamma_1, \gamma_2, \gamma_3) = \gamma_1 + \frac{\gamma_2}{\gamma_3 + \rho} \tag{10.7}$$

using standard nonlinear regression. Once we have estimates of the γ's, we take

$$\widetilde{\beta}_1 = G(-1; \widehat{\gamma}_1, \widehat{\gamma}_2, \widehat{\gamma}_3) \tag{10.8}$$

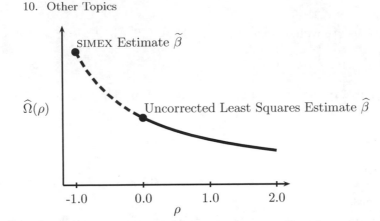

FIGURE 10.2. In the SIMEX method we extrapolate $\widehat{\Omega}(\rho)$ back to $\rho = -1$.

as our corrected estimate of β_1. Fitting the nonlinear regression (10.7) is inconvenient; it often suffices to approximate $G(\rho)$ with a quadratic. Thus, we fit the $\widehat{\Omega}(\rho_j)$ values to the curve

$$Q(\rho; \gamma_1, \gamma_2, \gamma_3) = \gamma_1 + \gamma_2 \rho + \gamma_3 \rho^2$$

and the corrected estimate of β_1 is

$$\widetilde{\beta}_1 = Q(-1; \widehat{\gamma}_1, \widehat{\gamma}_2, \widehat{\gamma}_3) = \widehat{\gamma}_1 - \widehat{\gamma}_2 + \widehat{\gamma}_3.$$

An advantage of SIMEX is that it extends readily to nonparametric regression. Let $\widehat{r}_n(x)$ be an uncorrected estimate of $r(x)$ obtained by regressing the Y_i's on the $X_i^\bullet$'s in the nonparametric problem

$$
\begin{aligned}
Y_i &= r(X_i) + \epsilon_i \\
X_i^\bullet &= X_i + U_i.
\end{aligned}
$$

Now perform the SIMEX algorithm to get $\widehat{r}_n(x, \rho)$ and define the corrected estimator $\widetilde{r}_n(x) = \widehat{r}_n(x, -1)$. There remains the problem of choosing the smoothing parameter. This is an area of active research; see Staudenmayer and Ruppert (2004), for example.

A more direct way to deal with measurement error is suggested by Fan and Truong (1993). They propose the kernel estimator

$$\widehat{r}_n(x) = \frac{\sum_{i=1}^{n} K_n \left(\frac{x - X_i^\bullet}{h_n} \right) Y_i}{\sum_{i=1}^{n} K_n \left(\frac{x - X_i^\bullet}{h_n} \right)} \tag{10.9}$$

where

$$K_n(x) = \frac{1}{2\pi} \int e^{-itx} \frac{\phi_K(t)}{\phi_U(t/h_n)} dt,$$

where ϕ_K is the Fourier transform of a kernel K and ϕ_U is the characteristic function of U. This is a standard kernel estimator except for the unusual kernel K_n, which will be motivated later in this section, after equation (10.21).

Yet another way to deal with measurement error, due to Stefanski (1985), is motivated by considering the asymptotics in which $\sigma_u \to 0$ rather than keeping σ_u fixed as n increases. Let us apply Stefanski's "small σ_u" approach to the nonparametric regression problem. Write the uncorrected estimator as

$$\widehat{r}_n(x) = \sum_{i=1}^{n} Y_i \, \ell_i(x, X_i^\bullet) \tag{10.10}$$

where we have written the weights as $\ell_i(x, X_i^\bullet)$ to emphasize the dependence on $X_i^\bullet$. If the X_i's had been observed, the estimator of r would be

$$r_n^*(x) = \sum_{i=1}^{n} Y_i \, \ell_i(x, X_i).$$

Expanding $\ell_i(x, X_i^\bullet)$ around X_i we have

$$\widehat{r}_n(x) \approx r_n^*(x) + \sum_{i=1}^{n} Y_i (X_i^\bullet - X_i)\ell'(x, X_i) + \frac{1}{2}\sum_{i=1}^{n} Y_i (X_i^\bullet - X_i)^2 \ell''(x, X_i). \tag{10.11}$$

Taking expectations, we see that the excess bias due to measurement error (conditional on the X_is) is

$$b(x) = \frac{\sigma_u^2}{2}\sum_{i=1}^{n} r(X_i)\ell''(x, X_i). \tag{10.12}$$

We can estimate $b(x)$ with

$$\widehat{b}(x) = \frac{\sigma_u^2}{2}\sum_{i=1}^{n} \widehat{r}(X_i^\bullet)\ell''(x, X_i^\bullet). \tag{10.13}$$

This yields a bias corrected estimate for r, namely,

$$\widetilde{r}_n(x) = \widehat{r}_n(x) - \widehat{b}(x) = \widehat{r}_n(x) - \frac{\sigma_u^2}{2}\sum_{i=1}^{n} \widehat{r}(X_i^\bullet)\ell''(x, X_i^\bullet). \tag{10.14}$$

This estimator still has measurement error bias due to the estimation of $b(x)$ but, for small σ_u^2, it will be less biased than $\widehat{r}_n$.

Now consider density estimation. Suppose that

$$X_1, \ldots, X_n \;\sim\; F$$
$$X_i^\bullet \;=\; X_i + U_i, \quad i = 1, \ldots, n$$

where, as before, the X_i's are not observed. We want to estimate the density $f(x) = F'(x)$. The density of $X^{\bullet}$ is

$$f^{\bullet}(x) = \int f(s) f_U(x - s) ds \qquad (10.15)$$

where f_U is the density of U. Since $f^{\bullet}$ is the convolution of f and f_U, the problem of estimating f is called **deconvolution**.

One way to estimate f is to use Fourier inversion. Let $\psi(t) = \int e^{itx} f(x) dx$ denote the Fourier transform (characteristic function) of X and define $\psi^{\bullet}$ and ψ_U similarly. Since $X^{\bullet} = X + U$, it follows that $\psi^{\bullet}(t) = \psi(t)\psi_U(t)$ and hence

$$\psi(t) = \frac{\psi^{\bullet}(t)}{\psi_U(t)}. \qquad (10.16)$$

If $\widehat{f}^{\bullet}$ is an estimate of $f^{\bullet}$ then

$$\widehat{\psi}^{\bullet}(t) = \int e^{itx} \widehat{f}^{\bullet}(x) dx \qquad (10.17)$$

is an estimate of $\psi^{\bullet}$. By Fourier inversion and equation (10.16),

$$f(x) = \frac{1}{2\pi} \int e^{-itx} \psi(t) dt = \frac{1}{2\pi} \int e^{-itx} \frac{\psi^{\bullet}(t)}{\psi_U(t)} dt \qquad (10.18)$$

which suggests the estimator

$$\widehat{f}(x) = \frac{1}{2\pi} \int e^{-itx} \frac{\widehat{\psi}^{\bullet}(t)}{\psi_U(t)} dt. \qquad (10.19)$$

In particular, if $\widehat{f}^{\bullet}$ is a kernel estimator,

$$\widehat{f}^{\bullet}(x) = \frac{1}{nh} \sum_{j=1}^{n} K\left(\frac{x - X_j^{\bullet}}{h}\right)$$

then $\widehat{f}(x)$ can be written as

$$\widehat{f}(x) = \frac{1}{nh} \sum_{j=1}^{n} K_*\left(\frac{x - X_j^{\bullet}}{h}, h\right) \qquad (10.20)$$

where

$$K_*(t, h) = \frac{1}{2\pi} \int e^{itu} \frac{\psi_K(u)}{\psi_U(u/h)} du \qquad (10.21)$$

and ψ_K is the Fourier transform of the kernel K. Equations (10.20) and (10.21) motivate the kernel regression estimator (10.9).

The risk of (10.20) is

$$\mathbb{E}(\widehat{f}_Y(y) - f(y))^2 \approx ch^4 + \frac{1}{2\pi nh} \int \left(\frac{\psi_K(t)}{|\psi_U(t/h)|} \right)^2 dt \qquad (10.22)$$

where

$$c = \frac{1}{4} \int y^2 K(y) dy \int (f''(y))^2 dy. \qquad (10.23)$$

Note that $\psi_U(t/h)$ appears in the denominator of (10.22). Thus, if $\psi_U(t/h)$ has thin tails, the risk will be large. Now $\psi_U(t/h)$ has thin tails when f_U is smooth, implying that if f_U is smooth, the rate of convergence is slow. In particular, if f_U is Normal, it can be shown that the best rate of convergence is $O(1/(\log n))^2$ which is very slow. Stefanski (1990) showed the surprising result that, under fairly general conditions, the optimal bandwidth is $h = \sigma_u/\sqrt{\log n}$ which does not depend on f.

In these asymptotic calculations, n is increasing while $\sigma_u^2 = \mathbb{V}(U)$ stays fixed. As we mentioned earlier, a more realistic asymptotic calculation might also have σ_u^2 tending to 0. In this approach, the convergence rate is less discouraging. The small σ_u approach suggests the corrected estimator

$$\widehat{f}_n(x) - \frac{\sigma_u^2}{2nh} \sum_{i=1}^{n} K'' \left(\frac{X_j^{\bullet} - x}{h} \right)$$

where $\widehat{f}_n$ is the naive, uncorrected kernel estimator using the $X_i^{\bullet}$s.

10.2   Inverse Problems

A class of problems that are very similar to measurement error are **inverse problems**. In general, an inverse problem refers to reconstructing properties of an object given only partial information on an object. An example is trying to estimate the three-dimensional structure of an object given information from two-dimensional slices of that object. This is common in certain types of medical diagnostic procedures. Another example is image reconstruction when the image is subject to blurring. Our summary here follows O'Sullivan (1986).

In the regression framework, a common inverse problem is of the form

$$Y_i = T_i(r) + \epsilon_i, \quad i = 1, \dots, n \qquad (10.24)$$

where r is the regression function of interest and T_i is some operator acting on r. A concrete example, which we shall use for the rest of this section, is $T_i(r) =$

$\int K_i(s)r(s)ds$ for some smooth function K_i such as $K_i(s) = e^{-(s-x_i)^2/2}$. The model becomes

$$Y_i = \int K_i(s)r(s)ds + \epsilon_i. \tag{10.25}$$

If K_i is a delta function at x_i, then (10.25) becomes the usual nonparametric regression model $Y_i = r(x_i) + \epsilon_i$. Think of $\int K_i(s)r(s)ds$ as a blurred version of r. There are now two types of information loss: the noise ϵ_i and blurring.

Suppose we estimate r with a linear smoother as defined in Section 5.2:

$$\widehat{r}_n(x) = \sum_{i=1}^{n} Y_i \ell_i(x). \tag{10.26}$$

The variance is the same as without blurring, namely, $\mathbb{V}(\widehat{r}_n(x)) = \sigma^2 \sum_{i=1}^{n} \ell_i^2(x)$ but the mean has a different form:

$$\mathbb{E}(\widehat{r}_n(x)) = \sum_{i=1}^{n} \ell_i(x) \int K_i(s)r(s)ds = \int A(x, s)r(s)ds$$

where

$$A(x, s) = \sum_{i=1}^{n} \ell_i(x)K_i(s) \tag{10.27}$$

is called the **Backus–Gilbert averaging kernel**.

Suppose that r can be approximated as an expansion in some basis $\{\phi_1, \ldots, \phi_k\}$ (see Chapter 8), that is, $r(x) = \sum_{j=1}^{k} \theta_j \phi_j(x)$. Then,

$$\int K_i(s)r(s)ds = \int K_i(s) \sum_{j=1}^{k} \theta_j \phi_j(s)ds = Z_i^T \theta$$

where $\theta = (\theta_1, \ldots, \theta_k)^T$ and

$$Z_i = \begin{pmatrix} \int K_i(s)\phi_1(s)ds \\ \int K_i(s)\phi_2(s)ds \\ \vdots \\ \int K_i(s)\phi_k(s)ds \end{pmatrix}.$$

The model (10.25) can then be written as

$$Y = Z\theta + \epsilon \tag{10.28}$$

where Z is an $n \times k$ matrix with i^{th} row equal to Z_i^T, $Y = (Y_1, \ldots, Y_n)^T$ and $\epsilon = (\epsilon_1, \ldots, \epsilon_n)^T$. It is tempting to estimate θ by the least squares estimator $(Z^T Z)^{-1}Z^T Y$. This may fail since $Z^T Z$ is typically not invertible in which case the problem is said to be **ill-posed**. Indeed, this is a hallmark of inverse problems and corresponds to the fact that the function r cannot be recovered, even in the absence of noise, due to the information loss incurred by blurring. Instead, it is common to use a regularized estimator such as $\widehat{\theta} = LY$ where

$$L = (Z^T Z + \lambda I)^{-1} Z^T,$$

I is the identity matrix and $\lambda > 0$ is a smoothing parameter that can be chosen by cross-validation. It should be noted that cross-validation is estimating the prediction error $n^{-1} \sum_{i=1}^{n} (\int K_i(s) r(s) ds - \int K_i(s) \widehat{r}(s) ds)^2$ rather than $\int (r(s) - \widehat{r}(s))^2 ds$. In Chapter 5 we observed that these two loss functions were essentially the same. This is no longer true in the present context. In principle, it is also possible to devise a cross-validation estimator for the loss $\int (r(s) - \widehat{r}(s))^2 ds$ but this estimator may be very unstable.

10.3 Nonparametric Bayes

Throughout this book we have taken a frequentist approach to inference. It is also possible to take a Bayesian approach.[1] Indeed, Bayesian nonparametric inference is a thriving research enterprise in statistics as well as in machine learning. Good references are Ghosh and Ramamoorthi (2003), Dey et al. (1998), Walker (2004) and references therein. However, the area is far to large and growing far too quickly for us to cover the topic here.

A small sampling of relevant references, in addition to those already mentioned, includes Schwartz (1965), Diaconis and Freedman (1986), Barron et al. (1999b), Ghosal et al. (1999), Walker and Hjort (2001), Hjort (2003), Ghosal et al. (2000), Shen and Wasserman (2001), Zhao (2000), Huang (2004), Cox (1993), Freedman (1999), McAuliffe et al. (2004), Teh et al. (2004), Blei et al. (2004), Blei and Jordan (2004), and Wasserman (1998).

10.4 Semiparametric Inference

As the name suggests, **semiparametric models** are models that are part parametric and part nonparametric. An example is a partially linear regression model of the form

$$Y = \beta X + r(Z) + \epsilon \qquad (10.29)$$

---

[1] See Chapter 11 of Wasserman (2004) for a general discussion of the advantages and pitfalls of Bayesian inference.

where r is some smooth function. The theory of inference for such models can get quite complex. Consider estimating β in (10.29), for example. One strategy is to regress $Y_i - \widehat{r}_n(Z_i)$ on X_i where $\widehat{r}_n$ is an estimate of r. Under appropriate conditions, if $\widehat{r}_n$ is chosen carefully, this will lead to good estimates of β. See Bickel et al. (1993) and Chapter 25 of van der Vaart (1998) for details.

10.5 Correlated Errors

If the errors ϵ_i in the regression model $Y_i = r(x_i) + \epsilon_i$ are correlated, then the usual methods can break down. In particular, positive correlation can make methods like cross-validation to choose very small bandwidths. There are several approaches for dealing with the correlation. In **modified cross-validation**, we drop out blocks of observations instead of single observations. In **partitioned cross-validation**, we partition the data, and use one observation per partition to construct the cross-validation estimate. The estimates obtained this way are then averaged. The properties of these methods are discussed in Chu and Marron (1991). A review of nonparametric regression with correlated observations can be found in Opsomer et al. (2001).

10.6 Classification

In the **classification problem** we have data $(X_1, Y_1), \ldots, (X_n, Y_n)$ where Y_i is discrete. We want to find a function $\widehat{h}$ so that, given a new X, we can predict Y by $\widehat{h}(X)$. This is just like regression except for two things: (i) the outcome is discrete and (ii) we do not need to estimate the relationship between Y and X well; we only need to predict well.

In an earlier version of this book, there was a long chapter on classification. The topic is so vast that the chapter took on a life of its own and I decided to eliminate it. There are many good books on classification such as Hastie et al. (2001). I will only make a few brief comments here.

Suppose that $Y_i \in \{0, 1\}$ is binary. A classifier is a function h that maps each x into $\{0, 1\}$. A commonly used risk function for classification is $L(h) = \mathbb{P}(Y \neq h(X))$. It can be shown that the optimal classification rule—called the

Bayes rule[2]—is

$$h(x) = \begin{cases} 1 & \text{if } r(x) \geq 1/2 \\ 0 & \text{if } r(x) < 1/2 \end{cases}$$

where $r(x) = \mathbb{E}(Y|X = x)$. This suggests a natural (and not uncommon) approach to classification. Form an estimate $\widehat{r}_n(x)$ of $r(x)$ then estimate h by

$$\widehat{h}(x) = \begin{cases} 1 & \text{if } \widehat{r}_n(x) \geq 1/2 \\ 0 & \text{if } \widehat{r}_n(x) < 1/2. \end{cases}$$

Now even if $\widehat{r}_n$ is a poor estimator of r, $\widehat{h}$ might still be a good classifier. For example, if $r(x) = .6$ but $\widehat{r}_n(x) = .9$, we still have that $h(x) = \widehat{h}(x) = 1$.

10.7 Sieves

A **sieve** is a sequence of models, indexed by sample size n, that increase in complexity as $n \to \infty$. A simple example is polynomial regression where the maximum degree of the polynomial $p(n)$ increases with n. Choosing $p(n)$ is like choosing a bandwidth: there is the usual tradeoff between bias and variance.

We use sieves informally all the time in the sense that we often fit more complex models when we have more data. The sieve idea was made formal by Grenander (1981) and Geman and Hwang (1982). Since then an enormous literature has developed. See Shen et al. (1999), Wong and Shen (1995), Shen and Wong (1994), Barron et al. (1999a), van de Geer (1995), Genovese and Wasserman (2000), and van de Geer (2000).

10.8 Shape-Restricted Inference

In the presence of shape restrictions, it is possible to make consistent, non-parametric inferences for a curve without imposing smoothness constraints. A typical example is estimating a regression function r when r is monotonic. A standard reference is Robertson et al. (1988).

Suppose that

$$Y_i = r(x_i) + \epsilon_i, \quad i = 1, \ldots, n$$

where $x_1 < \cdots < x_n$, $\mathbb{E}(\epsilon_i) = 0$ and $\sigma^2 = \mathbb{E}(\epsilon_i^2)$. Further, suppose that r is nondecreasing. (This assumption can be tested as described in Section

---

[2]This is a poor choice of terminology. The Bayes rule has nothing to do with Bayesian statistical inference. Indeed, the Bayes rule h can be estimated by frequentist or Bayesian methods.

10.9.) The least squares estimator $\widehat{r}_n$ is obtained by solving the restricted minimization:

$$\text{minimize } \sum_{i=1}^{n}(Y_i - r(x_i))^2 \quad \text{subject to } r \in \mathcal{F}_\uparrow$$

where $\mathcal{F}_\uparrow$ is the set of nondecreasing functions. The resulting estimator $\widehat{r}_n$ is called the **isotonic regression estimator**.

The solution $\widehat{r}_n$ may be described as follows. Let $P_0 = (0,0)$ and $P_j = (j, \sum_{i=1}^{j} Y_i)$. Let $G(t)$ be the **greatest convex minorant**, that is, $G(t)$ is the supremum of all convex functions that lie below the points $P_0, \ldots, P_n$. Then $\widehat{r}_n$ is the left derivative of G.

The convex minorant G can be found quickly using the pooled-adjacent-violators (PAV) algorithm. Start by joining all the points $P_0, P_1, \ldots$ with line segments. If the slope between P_0 and P_1 is greater than the slope between P_1 and P_2, replace these two segments with one line segment joining P_0 and P_2. If the slope between P_0 and P_2 is greater than the slope between P_2 and P_3, replace these two segments with one line segment joining P_0 and P_3. The process is continued in this way and the result is the function $G(t)$. See pages 8–10 of Robertson et al. (1988) for more detail.

A number of results are available about the resulting estimator. For example, Zhang (2002) shows the following. If

$$R_{n,p}(r) = \left(\frac{1}{n} \sum_{i=1}^{n} \mathbb{E}|\widehat{r}_n(x_i) - r(x_i)|^p \right)^{1/p}$$

where $1 \leq p \leq 3$, then

$$0.64 + o(1) \leq \frac{n^{1/3}}{\sigma^{2/3} V^{1/3}} \sup_{V(r) \leq V} R_{n,p}(r) \leq M_p + o(1) \tag{10.30}$$

where $V(r)$ is the total variation of r and M_p is a constant.

Optimal confidence bands are obtained in Dümbgen (2003), and Dümbgen and Johns (2004). Confidence bands for monotone densities are obtained in Hengartner and Stark (1995).

10.9   Testing

In this book we have concentrated on estimation and confidence sets. There is also a large literature on nonparametric testing. Many of the results can be found in the monograph by Ingster and Suslina (2003). Other references include: Ingster (2002), Ingster (2001), Ingster and Suslina (2000), Ingster (1998), Ingster (1993a), Ingster (1993b), Ingster (1993c), Lepski and Spokoiny (1999), and Baraud (2002).

For example, let $Y_i = \theta_i + \epsilon_i$ where $\epsilon_i \sim N(0,1)$, $i = 1,\ldots,n$ and $\theta = (\theta_1,\ldots,\theta_n)$ is unknown. Consider testing

$$H_0 : \theta = (0,\ldots,0) \quad \text{versus} \quad H_1 : \theta \in V_n = \left\{\theta \in \mathbb{R}^n : \|\theta\|_p \geq R_n\right\}$$

where

$$\|\theta\|_p = \left(\sum_{i=1}^{n} |\theta_i|^p\right)^{1/p}$$

for $0 < p < \infty$. The type I and type II error of a test ψ are

$$\alpha_n(\psi) = \mathbb{E}_0(\psi), \quad \beta_n(\psi,\theta) = \mathbb{E}_\theta(1 - \psi).$$

Let

$$\gamma_n = \inf_\psi \left(\alpha_n(\psi) + \sup_{\theta \in V_n} \beta_n(\psi,\theta)\right)$$

be the smallest possible sum of the type I error and maximum type II error, over all tests. Ingster showed that

$$\gamma_n \to 0 \iff \frac{R_n}{R_n^*} \to \infty \quad \text{and} \quad \gamma_n \to 1 \iff \frac{R_n}{R_n^*} \to 0$$

where

$$R_n^* = \begin{cases} n^{(1/p)-(1/4)} & \text{if } p \leq 2 \\ n^{1/(2p)} & \text{if } p > 2. \end{cases}$$

Thus, R_n^* is a critical rate that determines when the alternative is distinguishable. Results like these are intimately related to the results on confidence sets in Chapter 7.

The results for qualitative hypotheses are of a different nature. These are hypotheses like: f is monotone, f is positive, f is convex and so on. The defining characteristic of such hypotheses is that they are cones, that is, they are closed under addition. For example, if f and g are monotone nondecreasing functions, then $f + g$ is also monotone nondecreasing. References include: Dümbgen and Spokoiny (2001), Baraud et al. (2003a), Baraud et al. (2003b), and Juditsky and Nemirovski (2002). Consider testing the null hypothesis that a regression function r is nonincreasing. Suppose further that

$$r \in \left\{f : [0,1] \to \mathbb{R} : |r(x) - r(y)| \leq L|x - y|^s, \text{ for all } x, y \in [0,1]\right\}$$

where $L > 0$ and $0 < s \leq 1$. Then there exist tests with uniformly large power for each function whose distance from the null is at least of order $L^{1/(1+2s)} n^{-s/(1+2s)}$.

10.10   Computational Issues

I have completely neglected issues about efficient computation. Nonparametric methods work best with large data sets but implementing nonparametric methods with large data sets requires efficient computation.

Binning methods are quite popular for fast calculations. See Hall and Wand (1996), Fan and Marron (1994), Wand (1994), Holmström (2000), Sain (2002), and Scott (1992), for example. Chapter 12 of Loader (1999a) contains a good discussion of computation. In particular, there is a description of **k-d trees** which are intelligently chosen partitions of the data that speed up computations. A number of publications on the use of k-d trees in statistics can be found at `http://www.autonlab.org`. Useful R code can be found at `http://cran.r-project.org`. The `locfit` program by Catherine Loader for local likelihood and local regression can be found at `http://www.locfit.info`.

10.11   Exercises

1. Consider the "errors in Y" model:

$$
\begin{aligned}
Y_i &= r(X_i) + \epsilon_i \\
Y_i^\bullet &= Y_i + U_i, \qquad \mathbb{E}(U_i) = 0
\end{aligned}
$$

and the observed data are $(X_1, Y_1^\bullet), \ldots, (X_n, Y_n^\bullet)$. How does observing $Y_i^\bullet$ instead of Y_i affect the estimation of $r(x)$?

2. Prove 10.1.

3. Prove equation (10.20).

4. Draw $X_1, \ldots, X_n \sim N(0, 1)$ with $n = 100$.

    (a) Estimate the density using a kernel estimator.

    (b) Let $W_i = X_i + \sigma_u U_i$ where $U_i \sim N(0, 1)$. Compute the uncorrected and corrected density estimator from the W_is and compare the estimators. Try different values of σ_u.

    (c) Repeat part (b) but let U_i have a Cauchy distribution.

5. Generate 1000 observations from the model:

$$
\begin{aligned}
Y_i &= r(X_i) + \sigma_\epsilon \epsilon_i \\
W_i &= X_i + \sigma_u U_i
\end{aligned}
$$

where $r(x) = x + 3\exp(-16x^2)$, $\epsilon_i \sim N(0,1)$, $U_i \sim N(0,1)$, $X_i \sim$ Unif$(-2,2)$, $\sigma_\epsilon = 0.5$, and $\sigma_u = 0.1$.

(a) Use kernel regression to estimate r from $(X_1, Y_1), \ldots, (X_n, Y_n)$. Find the bandwidth h using cross-validation. Call the resulting estimator r_n^*. Use the bandwidth h throughout the rest of the question.

(b) Use kernel regression to estimate r from $(W_1, Y_1), \ldots, (W_n, Y_n)$. Denote the resulting estimator by $\widehat{r}_n$.

(c) Compute the corrected estimator $\widetilde{r}_n$ given by (10.14).

(d) Compare r, r_n^*, $\widehat{r}_n$ and $\widetilde{r}_n$.

(e) Think of a way of finding a good bandwidth using only the Y_i's and the W_i's. Implement your method and compare the resulting estimator to the previous estimators.

6. Generate 1000 observations from the model:

$$Y_i = \int K_i(s)ds + \sigma\epsilon_i$$

where $r(x) = x + 3\exp(-16x^2)$, $\epsilon_i \sim N(0,1)$, $\sigma = 0.5$, $K_i(s) = e^{-(s-x_i)^2/b^2}$, and $x_i = 4(i/n) - 2$. Try $b = .01, .1$ and 1.

(a) Plot the Backus–Gilbert averaging kernel for several values of x. Interpret.

(b) Estimate r using the method described in this chapter. Comment on the results.

7. Consider the infinite-dimensional Normal means model from Chapter 7 given by:

$$Y_i = \theta_i + \frac{1}{\sqrt{n}}\epsilon_i, \quad i = 1, 2, \ldots$$

Assume that $\theta = (\theta_1, \theta_2, \ldots)$ is in the Sobolev ellipsoid $\Theta = \{\theta : \sum_{j=1}^\infty \theta_j^2 j^2 \leq c^2\}$ for some $c > 0$. Let the prior be such that each θ_i is independent and $\theta_i \sim N(0, \tau_i^2)$ for some $\tau_i > 0$.

(a) Find the posterior of the θ_i's. In particular, find the posterior mean $\widehat{\theta}$.

(b) Find conditions on the τ_i^2 such that the posterior is consistent in the sense that

$$\mathbb{P}_\theta(\|\theta - \widehat{\theta}\| > \epsilon) \to 0$$

for any ϵ, as $n \to \infty$.

(c) Let $\theta_j = 1/j^4$. Simulate from the model (take $n = 100$), find the posterior and find b_n such that $\mathbb{P}(||\widehat{\theta} - \theta|| \leq b_n | \text{Data}) = 0.95$. Set $B_n = \{\theta \in \Theta : ||\widehat{\theta} - \theta|| \leq b_n\}$. Repeat this whole process many times and estimate the frequentist coverage of B_n (at this particular θ) by counting how often $\theta \in B_n$. Report your findings.

(d) Repeat (c) for $\theta = (0, 0, \ldots)$.

Bibliography

AKAIKE, H. (1973). Information theory and an extension of the maximum likelihood principle. *Second International Symposium on Information Theory* 267–281.

BARAUD, Y. (2002). Nonasymptotic minimax rates of testing in signal detection. *Bernoulli* **8** 577–606.

BARAUD, Y. (2004). Confidence balls in Gaussian regression. *The Annals of Statistics* **32** 528–551.

BARAUD, Y., HUET, S. and LAURENT, B. (2003a). Adaptive tests of linear hypotheses by model selection. *The Annals of Statistics* **31** 225–251.

BARAUD, Y., HUET, S. and LAURENT, B. (2003b). Adaptive tests of qualitative hypotheses. *ESAIM P&S: Probability and Statistics* **7** 147–159.

BARRON, A., BIRGE, L. and MASSART, P. (1999a). Risk bounds for model selection via penalization. *Probability Theory and Related Fields* **113** 301–413.

BARRON, A., SCHERVISH, M. J. and WASSERMAN, L. (1999b). The consistency of posterior distributions in nonparametric problems. *The Annals of Statistics* **27** 536–561.

BELLMAN, R. (1961). *Adaptive Control Processes*. Princeton University Press. Princeton, NJ.

BERAN, R. (2000). REACT scatterplot smoothers: Superefficiency through basis economy. *Journal of the American Statistical Association* **95** 155–171.

BERAN, R. and DÜMBGEN, L. (1998). Modulation of estimators and confidence sets. *The Annals of Statistics* **26** 1826–1856.

BICKEL, P. J. and FREEDMAN, D. A. (1981). Some asymptotic theory for the bootstrap. *The Annals of Statistics* **9** 1196–1217.

BICKEL, P. J., KLAASSEN, C. A. J., RITOV, Y. and WELLNER, J. A. (1993). *Efficient and adaptive estimation for semiparametric models*. Johns Hopkins University Press. Baltimore, MD.

BLEI, D., GRIFFITHS, T., JORDAN, M. and TENEBAUM, J. (2004). Hierarchical topic models and the nested Chinese restaurant process. *In S. Thrun, L. Saul, and B. Schoelkopf (Eds.), Advances in Neural Information Processing Systems (NIPS) 16, 2004.* .

BLEI, D. and JORDAN, M. (2004). Variational methods for the dirichlet process. In *Proceedings of the 21st International Conference on Machine Learning (ICML)*. Omnipress.

BREIMAN, L., FRIEDMAN, J. H., OLSHEN, R. A. and STONE, C. J. (1984). *Classification and regression trees*. Wadsworth. New York, NY.

BROWN, L., CAI, T. and ZHOU, H. (2005). A root-unroot transform and wavelet block thresholding approach to adaptive density estimation. *unpublished* .

BROWN, L. D. and LOW, M. G. (1996). Asymptotic equivalence of nonparametric regression and white noise. *The Annals of Statistics* **24** 2384–2398.

CAI, T. and LOW, M. (2005). Adaptive confidence balls. *To appear: The Annals of Statistics* .

CAI, T., LOW, M. and ZHAO, L. (2000). Sharp adaptive estimation by a blockwise method. *Technical report, Wharton School, University of Pennsylvania, Philadelphia* .

CARROLL, R., RUPPERT, D. and STEFANSKI, L. (1995). *Measurement Error in Nonlinear Models*. Chapman and Hall. New York, NY.

CASELLA, G. and BERGER, R. L. (2002). *Statistical Inference.* Duxbury Press. New York, NY.

CENČOV, N. (1962). Evaluation of an unknown distribution density from observations. *Doklady* **3** 1559–1562.

CHAUDHURI, P. and MARRON, J. S. (1999). Sizer for exploration of structures in curves. *Journal of the American Statistical Association* **94** 807–823.

CHAUDHURI, P. and MARRON, J. S. (2000). Scale space view of curve estimation. *The Annals of Statistics* **28** 408–428.

CHEN, S. S., DONOHO, D. L. and SAUNDERS, M. A. (1998). Atomic decomposition by basis pursuit. *SIAM Journal on Scientific Computing* **20** 33–61.

CHEN, S. X. and QIN, Y. S. (2000). Empirical likelihood confidence intervals for local linear smoothers. *Biometrika* **87** 946–953.

CHEN, S. X. and QIN, Y. S. (2002). Confidence intervals based on local linear smoother. *Scandinavian Journal of Statistics* **29** 89–99.

CHU, C.-K. and MARRON, J. S. (1991). Comparison of two bandwidth selectors with dependent errors. *The Annals of Statistics* **19** 1906–1918.

CLAESKENS, G. and HJORT, N. (2004). Goodness-of-fit via nonparametric likelihood ratios. *Scandinavian Journal of Statistics* **31** 487–513.

COOK, J. R. and STEFANSKI, L. A. (1994). Simulation-extrapolation estimation in parametric measurement error models. *Journal of the American Statistical Association* **89** 1314–1328.

COX, D. and LEWIS, P. (1966). *The Statistical Analysis of Series of Events.* Chapman and Hall. New York, NY.

COX, D. D. (1993). An analysis of Bayesian inference for nonparametric regression. *The Annals of Statistics* **21** 903–923.

CUMMINS, D. J., FILLOON, T. G. and NYCHKA, D. (2001). Confidence intervals for nonparametric curve estimates: Toward more uniform pointwise coverage. *Journal of the American Statistical Association* **96** 233–246.

DAUBECHIES, I. (1992). *Ten Lectures on Wavelets.* SIAM. New York, NY.

DAVISON, A. C. and HINKLEY, D. V. (1997). *Bootstrap Methods and Their Application*. Cambridge University Press. Cambridge.

DEVROYE, L., GYÖRFI, L. and LUGOSI, G. (1996). *A Probabilistic Theory of Pattern Recognition*. Springer-Verlag. New York, NY.

DEY, D., MULLER, P. and SINHA, D. (1998). *Practical Nonparametric and Semiparametric Bayesian Statistics*. Springer-Verlag. New York, NY.

DIACONIS, P. and FREEDMAN, D. (1986). On inconsistent Bayes estimates of location. *The Annals of Statistics* **14** 68–87.

DONOHO, D. L. and JOHNSTONE, I. M. (1994). Ideal spatial adaptation by wavelet shrinkage. *Biometrika* **81** 425–455.

DONOHO, D. L. and JOHNSTONE, I. M. (1995). Adapting to unknown smoothness via wavelet shrinkage. *Journal of the American Statistical Association* **90** 1200–1224.

DONOHO, D. L. and JOHNSTONE, I. M. (1998). Minimax estimation via wavelet shrinkage. *The Annals of Statistics* **26** 879–921.

DONOHO, D. L., JOHNSTONE, I. M., KERKYACHARIAN, G. and PICARD, D. (1995). Wavelet shrinkage: Asymptopia? *Journal of the Royal Statistical Society, Series B, Methodological* **57** 301–337.

DÜMBGEN, L. (2003). Optimal confidence bands for shape-restricted curves. *Bernoulli* **9** 423–449.

DÜMBGEN, L. and JOHNS, R. (2004). Confidence bands for isotonic median curves using sign-tests. *Journal of Computational and Graphical Statistics* **13** 519–533.

DÜMBGEN, L. and SPOKOINY, V. G. (2001). Multiscale testing of qualitative hypotheses. *The Annals of Statistics* **29** 124–152.

EFROMOVICH, S. (1999). *Nonparametric Curve Estimation: Methods, Theory and Applications*. Springer-Verlag. New York, NY.

EFROMOVICH, S. Y. and PINSKER, M. S. (1982). Estimation of square-integrable probability density of a random variable. *Problems of Information Transmission, (Transl of Problemy Peredachi Informatsii)* **18** 175–189.

EFROMOVICH, S. Y. and PINSKER, M. S. (1984). A learning algorithm for nonparametric filtering. *Automat. i Telemekh* **11** 58–65.

EFRON, B. (1979). Bootstrap methods: Another look at the jackknife. *The Annals of Statistics* **7** 1–26.

EFRON, B., HASTIE, T., JOHNSTONE, I. and TIBSHIRANI, R. (2004). Least angle regression. *The Annals of Statistics* **32** 407–499.

EFRON, B. and TIBSHIRANI, R. J. (1993). *An Introduction to the Bootstrap.* Chapman and Hall. New York, NY.

FAN, J. (1992). Design-adaptive nonparametric regression. *Journal of the American Statistical Association* **87** 998–1004.

FAN, J. and GIJBELS, I. (1995). Data-driven bandwidth selection in local polynomial fitting: Variable bandwidth and spatial adaptation. *Journal of the Royal Statistical Society, Series B, Methodological* **57** 371–394.

FAN, J. and GIJBELS, I. (1996). *Local Polynomial Modelling and Its Applications.* Chapman and Hall. New York, NY.

FAN, J. and MARRON, J. S. (1994). Fast implementations of nonparametric curve estimators. *Journal of Computational and Graphical Statistics* **3** 35–56.

FAN, J. and TRUONG, Y. K. (1993). Nonparametric regression with errors in variables. *The Annals of Statistics* **21** 1900–1925.

FARAWAY, J. J. (1990). Bootstrap selection of bandwidth and confidence bands for nonparametric regression. *Journal of Statistical Computation and Simulation* **37** 37–44.

FARAWAY, J. J. and SUN, J. (1995). Simultaneous confidence bands for linear regression with heteroscedastic errors. *Journal of the American Statistical Association* **90** 1094–1098.

FERNHOLZ, L. T. (1983). *Von Mises' Calculus for Statistical Functionals.* Springer-Verlag. New York, NY.

FREEDMAN, D. (1999). Wald lecture: On the Bernstein–von Mises theorem with infinite-dimensional parameters. *The Annals of Statistics* **27** 1119–1141.

FRIEDMAN, J. H. (1991). Multivariate adaptive regression splines. *The Annals of Statistics* **19** 1–67.

FRIEDMAN, J. H. and STUETZLE, W. (1981). Projection pursuit regression. *Journal of the American Statistical Association* **76** 817–823.

FULLER, W. A. (1987). *Measurement Error Models*. John Wiley. New York, NY.

GAO, F., WAHBA, G., KLEIN, R. and KLEIN, B. (2001). Smoothing spline ANOVA for multivariate Bernoulli observations with application to ophthalmology data. *Journal of the American Statistical Association* **96** 127–160.

GASSER, T., SROKA, L. and JENNEN-STEINMETZ, C. (1986). Residual variance and residual pattern in nonlinear regression. *Biometrika* **73** 625–633.

GEMAN, S. and HWANG, C.-R. (1982). Nonparametric maximum likelihood estimation by the method of sieves. *The Annals of Statistics* **10** 401–414.

GENOVESE, C., MILLER, C., NICHOL, R., ARJUNWADKAR, M. and WASSERMAN, L. (2004). Nonparametric inference for the cosmic microwave background. *Statistical Science* **19** 308–321.

GENOVESE, C. and WASSERMAN, L. (2005). Nonparametric confidence sets for wavelet regression. *Annals of Statistics* **33** 698–729.

GENOVESE, C. R. and WASSERMAN, L. (2000). Rates of convergence for the Gaussian mixture sieve. *The Annals of Statistics* **28** 1105–1127.

GHOSAL, S., GHOSH, J. K. and RAMAMOORTHI, R. V. (1999). Posterior consistency of Dirichlet mixtures in density estimation. *The Annals of Statistics* **27** 143–158.

GHOSAL, S., GHOSH, J. K. and VAN DER VAART, A. W. (2000). Convergence rates of posterior distributions. *The Annals of Statistics* **28** 500–531.

GHOSH, J. and RAMAMOORTHI, R. (2003). *Bayesian Nonparametrics*. Springer-Verlag. New York, NY.

GLAD, I., HJORT, N. and USHAKOV, N. (2003). Correction of density estimators that are not densities. *Scandinavian Journal of Statististics* **30** 415–427.

GOLDENSHLUGER, A. and NEMIROVSKI, A. (1997). On spatially adaptive estimation of nonparametric regression. *Mathematical Methods of Statistics* **6** 135–170.

GREEN, P. J. and SILVERMAN, B. W. (1994). *Nonparametric regression and generalized linear models: a roughness penalty approach.* Chapman and Hall. New York, NY.

GRENANDER, U. (1981). *Abstract Inference.* John Wiley. New York, NY.

HALL, P. (1987). On Kullback–Leibler loss and density estimation. *The Annals of Statistics* **15** 1491–1519.

HALL, P. (1992a). *The Bootstrap and Edgeworth Expansion.* Springer-Verlag. New York, NY.

HALL, P. (1992b). On bootstrap confidence intervals in nonparametric regression. *The Annals of Statistics* **20** 695–711.

HALL, P. (1993). On Edgeworth expansion and bootstrap confidence bands in nonparametric curve estimation. *Journal of the Royal Statistical Society, Series B, Methodological* **55** 291–304.

HALL, P. and WAND, M. P. (1996). On the accuracy of binned kernel density estimators. *Journal of Multivariate Analysis* **56** 165–184.

HÄRDLE, W. (1990). *Applied Nonparametric Regression.* Cambridge University Press. Cambridge.

HÄRDLE, W. and BOWMAN, A. W. (1988). Bootstrapping in nonparametric regression: Local adaptive smoothing and confidence bands. *Journal of the American Statistical Association* **83** 102–110.

HÄRDLE, W., HALL, P. and MARRON, J. S. (1988). How far are automatically chosen regression smoothing parameters from their optimum? *Journal of the American Statistical Association* **83** 86–95.

HÄRDLE, W., KERKYACHARIAN, G., PICARD, D. and TSYBAKOV, A. (1998). *Wavelets, Approximation, and Statistical Applications.* Springer-Verlag. New York, NY.

HÄRDLE, W. and MAMMEN, E. (1993). Comparing nonparametric versus parametric regression fits. *The Annals of Statistics* **21** 1926–1947.

HÄRDLE, W. and MARRON, J. S. (1991). Bootstrap simultaneous error bars for nonparametric regression. *The Annals of Statistics* **19** 778–796.

HASTIE, T. and LOADER, C. (1993). Local regression: Automatic kernel carpentry. *Statistical Science* **8** 120–129.

HASTIE, T. and TIBSHIRANI, R. (1999). *Generalized Additive Models.* Chapman and Hall. New York, NY.

HASTIE, T., TIBSHIRANI, R. and FRIEDMAN, J. H. (2001). *The Elements of Statistical Learning: Data Mining, Inference, and Prediction.* Springer-Verlag. New York, NY.

HENGARTNER, N. W. and STARK, P. B. (1995). Finite-sample confidence envelopes for shape-restricted densities. *The Annals of Statistics* **23** 525–550.

HJORT, N. (1999). Towards semiparametric bandwidth selectors for kernel density estimation. *Statistical Research Report.* Department of Mathematics, University of Oslo .

HJORT, N. (2003). Topics in nonparametric Bayesian statistics. In *Highly Structured Stochastic Systems. P. Green, N.L. Hjort, S. Richardson (Eds.).* Oxford University Press. Oxford.

HJORT, N. L. and JONES, M. C. (1996). Locally parametric nonparametric density estimation. *The Annals of Statistics* **24** 1619–1647.

HOLMSTRÖM, L. (2000). The accuracy and the computational complexity of a multivariate binned kernel density estimator. *Journal of Multivariate Analysis* **72** 264–309.

HOTELLING, H. (1939). Tubes and spheres in n-spaces, and a class of statistical problems. *American Journal of Mathematics* **61** 440–460.

HUANG, T.-M. (2004). Convergence rates for posterior distributions and adaptive estimation. *The Annals of Statistics* **32** 1556–1593.

IBRAGIMOV, I. A. and HAS'MINSKII, R. Z. (1977). On the estimation of an infinite-dimensional parameter in Gaussian white noise. *Soviet Math. Dokl.* **236** 1053–1055.

INGSTER, Y. and SUSLINA, I. (2003). *Nonparametric Goodness-of-Fit Testing Under Gaussian Models.* Springer-Verlag. New York, NY.

INGSTER, Y. I. (1993a). Asymptotically minimax hypothesis testing for nonparametric alternatives. I. *Mathematical Methods of Statistics* **2** 85–114.

INGSTER, Y. I. (1993b). Asymptotically minimax hypothesis testing for nonparametric alternatives. II. *Mathematical Methods of Statistics* **2** 171–189.

INGSTER, Y. I. (1993c). Asymptotically minimax hypothesis testing for nonparametric alternatives, III. *Mathematical Methods of Statistics* **2** 249–268.

INGSTER, Y. I. (1998). Minimax detection of a signal for l^n-balls. *Mathematical Methods of Statistics* **7** 401–428.

INGSTER, Y. I. (2001). Adaptive detection of a signal of growing dimension. I. *Mathematical Methods of Statistics* **10** 395–421.

INGSTER, Y. I. (2002). Adaptive detection of a signal of growing dimension. II. *Mathematical Methods of Statistics* **11** 37–68.

INGSTER, Y. I. and SUSLINA, I. A. (2000). Minimax nonparametric hypothesis testing for ellipsoids and Besov bodies. *ESAIM P&S: Probability and Statistics* **4** 53–135.

JANG, W., GENOVESE, C. and WASSERMAN, L. (2004). Nonparametric confidence sets for densities. *Technical Report,* Carnegie Mellon University, Pittsburgh .

JOHNSTONE, I. (2003). *Function Estimation in Gaussian Noise: Sequence Models.* Unpublished manuscript.

JUDITSKY, A. and LAMBERT-LACROIX, S. (2003). Nonparametric confidence set estimation. *Mathematical Methods of Statistics* **19** 410–428.

JUDITSKY, A. and NEMIROVSKI, A. (2002). On nonparametric tests of positivity/monotonicity/convexity. *The Annals of Statistics* **30** 498–527.

LEPSKI, O. (1999). How to improve the accuracy of estimation. *Mathematical Methods in Statistics* **8** 441–486.

LEPSKI, O. V., MAMMEN, E. and SPOKOINY, V. G. (1997). Optimal spatial adaptation to inhomogeneous smoothness: An approach based on kernel estimates with variable bandwidth selectors. *The Annals of Statistics* **25** 929–947.

LEPSKI, O. V. and SPOKOINY, V. G. (1999). Minimax nonparametric hypothesis testing: The case of an inhomogeneous alternative. *Bernoulli* **5** 333–358.

LEPSKII, O. V. (1991). On a problem of adaptive estimation in Gaussian white noise. *Theory of Probability and Its Applications (Transl of Teorija Verojatnostei i ee Primenenija)* **35** 454–466.

LI, K.-C. (1989). Honest confidence regions for nonparametric regression. *The Annals of Statistics* **17** 1001–1008.

LIN, X., WAHBA, G., XIANG, D., GAO, F., KLEIN, R. and KLEIN, B. (2000). Smoothing spline ANOVA models for large data sets with Bernoulli observations and the randomized GACV. *The Annals of Statistics* **28** 1570–1600.

LIN, Y. (2000). Tensor product space ANOVA models. *The Annals of Statistics* **28** 734–755.

LOADER, C. (1999a). *Local Regression and Likelihood*. Springer-Verlag. New York, NY.

LOADER, C. R. (1999b). Bandwidth selection: classical or plug-in? *The Annals of Statistics* **27** 415–438.

LOW, M. G. (1997). On nonparametric confidence intervals. *The Annals of Statistics* **25** 2547–2554.

MALLOWS, C. L. (1973). Some comments on C_p. *Technometrics* **15** 661–675.

MAMMEN, E. and VAN DE GEER, S. (1997). Locally adaptive regression splines. *The Annals of Statistics* **25** 387–413.

MARRON, J. S. and WAND, M. P. (1992). Exact mean integrated squared error. *The Annals of Statistics* **20** 712–736.

MCAULIFFE, J., BLEI, D. and JORDAN, M. (2004). *Variational inference for Dirichlet process mixtures*. Department of Statistics, University of California, Berkeley.

MCCULLAGH, P. and NELDER, J. A. (1999). *Generalized linear models*. Chapman and Hall. New York, NY.

MORGAN, J. N. and SONQUIST, J. A. (1963). Problems in the analysis of survey data, and a proposal. *Journal of the American Statistical Association* **58** 415–434.

MÜLLER, H.-G. and STADTMLLER, U. (1987). Variable bandwidth kernel estimators of regression curves. *The Annals of Statistics* **15** 182–201.

NAIMAN, D. Q. (1990). Volumes of tubular neighborhoods of spherical polyhedra and statistical inference. *The Annals of Statistics* **18** 685–716.

NEUMANN, M. H. (1995). Automatic bandwidth choice and confidence intervals in nonparametric regression. *The Annals of Statistics* **23** 1937–1959.

NEUMANN, M. H. and POLZEHL, J. (1998). Simultaneous bootstrap confidence bands in nonparametric regression. *Journal of Nonparametric Statistics* **9** 307–333

NUSSBAUM, M. (1985). Spline smoothing in regression models and asymptotic efficiency in L_2. *The Annals of Statistics* **13** 984–997.

NUSSBAUM, M. (1996a). Asymptotic equivalence of density estimation and Gaussian white noise. *The Annals of Statistics* **24** 2399–2430.

NUSSBAUM, M. (1996b). The Pinsker bound: A review. In *Encyclopedia of Statistical Sciences (S. Kotz, Ed)*. Wiley. New York, NY.

NYCHKA, D. (1988). Bayesian confidence intervals for smoothing splines. *Journal of the American Statistical Association* **83** 1134–1143.

OGDEN, R. T. (1997). *Essential Wavelets for Statistical Applications and Data Analysis*. Birkhäuser. Boston, MA.

OPSOMER, J., WANG, Y. and YANG, Y. (2001). Nonparametric regression with correlated errors. *Statistical Science* **16** 134–153.

O'SULLIVAN, F. (1986). A statistical perspective on ill-posed inverse problems. *Statistical Science* **1** 502–527.

PAGANO, M. and GAUVREAU, K. (1993). *Principles of biostatistics*. Duxbury Press. New York, NY.

PARZEN, E. (1962). On estimation of a probability density function and mode. *The Annals of Mathematical Statistics* **33** 1065–1076.

PICARD, D. and TRIBOULEY, K. (2000). Adaptive confidence interval for pointwise curve estimation. *The Annals of Statistics* **28** 298–335.

QUENOUILLE, M. (1949). Approximate tests of correlation in time series. *Journal of the Royal Statistical Society B* **11** 18–84.

RICE, J. (1984). Bandwidth choice for nonparametric regression. *The Annals of Statistics* **12** 1215–1230.

RICE, S. (1939). The distribution of the maxima of a random curve. *American Journal of Mathematics* **61** 409–416.

ROBERTSON, T., WRIGHT, F. T. and DYKSTRA, R. (1988). *Order restricted statistical inference.* Wiley. New York, NY.

ROBINS, J. and VAN DER VAART, A. (2005). Adaptive nonparametric confidence sets. *To appear: The Annals of Statistics* .

ROSENBLATT, M. (1956). Remarks on some nonparametric estimates of a density function. *Annals of Mathematical Statistics* **27** 832–837.

RUDEMO, M. (1982). Empirical choice of histograms and kernel density estimators. *Scandinavian Journal of Statistics* **9** 65–78.

RUPPERT, D. (1997). Empirical-bias bandwidths for local polynomial nonparametric regression and density estimation. *Journal of the American Statistical Association* **92** 1049–1062.

RUPPERT, D., WAND, M. and CARROLL, R. (2003). *Semiparametric Regression.* Cambridge University Press. Cambridge.

RUPPERT, D. and WAND, M. P. (1994). Multivariate locally weighted least squares regression. *The Annals of Statistics* **22** 1346–1370.

SAIN, S. R. (2002). Multivariate locally adaptive density estimation. *Computational Statistics and Data Analysis* **39** 165–186.

SCHEFFÉ, H. (1959). *The Analysis of Variance.* Wiley. New York, NY.

SCHWARTZ, L. (1965). On Bayes procedures. *Zeitschrift für Wahrscheinlichkeitstheorie und Verwandte Gebiete* **4** 10–26.

SCHWARZ, G. (1978). Estimating the dimension of a model. *The Annals of Statistics* **6** 461–464.

SCOTT, D. W. (1992). *Multivariate Density Estimation: Theory, Practice, and Visualization.* Wiley. New York, NY.

SERFLING, R. J. (1980). *Approximation Theorems of Mathematical Statistics.* Wiley. New York, NY.

SHAO, J. and TU, D. (1995). *The Jackknife and Bootstrap.* Springer-Verlag. New York, NY.

SHEN, X., SHI, J. and WONG, W. H. (1999). Random sieve likelihood and general regression models. *Journal of the American Statistical Association* **94** 835–846.

SHEN, X. and WASSERMAN, L. (2001). Rates of convergence of posterior distributions. *The Annals of Statistics* **29** 687–714.

SHEN, X. and WONG, W. H. (1994). Convergence rate of sieve estimates. *The Annals of Statistics* **22** 580–615.

SIGRIST, M. E. (1994). *Air Monitoring by Spectroscopic Techniques*. Wiley. New York, NY.

SILVERMAN, B. W. (1984). Spline smoothing: The equivalent variable kernel method. *The Annals of Statistics* **12** 898–916.

SILVERMAN, B. W. (1986). *Density Estimation for Statistics and Data Analysis*. Chapman and Hall. New York, NY.

SIMONOFF, J. S. (1996). *Smoothing Methods in Statistics*. Springer-Verlag. New York, NY.

SINGH, K. (1981). On the asymptotic accuracy of Efron's bootstrap. *The Annals of Statistics* **9** 1187–1195.

STAUDENMAYER, J. and RUPPERT, D. (2004). Local polynomial regression and simulation–extrapolation. *Journal of the Royal Statistical Society Series B* **66** 17–30.

STEFANSKI, L. A. (1985). The effects of measurement error on parameter estimation. *Biometrika* **72** 583–592.

STEFANSKI, L. A. (1990). Rates of convergence of some estimators in a class of deconvolution problems. *Statistics and Probability Letters* **9** 229–235.

STEFANSKI, L. A. and COOK, J. R. (1995). Simulation–extrapolation: The measurement error jackknife. *Journal of the American Statistical Association* **90** 1247–1256.

STEIN, C. M. (1981). Estimation of the mean of a multivariate normal distribution. *The Annals of Statistics* **9** 1135–1151.

STONE, C. J. (1984). An asymptotically optimal window selection rule for kernel density estimates. *The Annals of Statistics* **12** 1285–1297.

SUN, J. and LOADER, C. R. (1994). Simultaneous confidence bands for linear regression and smoothing. *The Annals of Statistics* **22** 1328–1345.

TEH, Y., JORDAN, M., BEAL, M. and BLEI, D. (2004). Hierarchical dirichlet processes. In *Technical Report*. Department of Statistics, University of California, Berkeley.

TIBSHIRANI, R. (1996). Regression shrinkage and selection via the lasso. *Journal of the Royal Statistical Society, Series B, Methodological* **58** 267–288.

TUKEY, J. (1958). Bias and confidence in not quite large samples. *The Annals of Mathematical Statistics* **29** 614.

VAN DE GEER, S. (1995). The method of sieves and minimum contrast estimators. *Mathematical Methods of Statistics* **4** 20–38.

VAN DE GEER, S. A. (2000). *Empirical Processes in M-Estimation*. Cambridge University Press. Cambridge.

VAN DER VAART, A. W. (1998). *Asymptotic Statistics*. Cambridge University Press.

VAN DER VAART, A. W. and WELLNER, J. A. (1996). *Weak Convergence and Empirical Processes: With Applications to Statistics*. Springer-Verlag.

VENABLES, W. N. and RIPLEY, B. D. (2002). *Modern Applied Statistics with S*. Springer-Verlag. New York, NY.

WAHBA, G. (1983). Bayesian "confidence intervals" for the cross-validated smoothing spline. *Journal of the Royal Statistical Society, Series B, Methodological* **45** 133–150.

WAHBA, G. (1990). *Spline models for observational data*. SIAM. New York, NY.

WAHBA, G., WANG, Y., GU, C., KLEIN, R. and KLEIN, B. (1995). Smoothing spline ANOVA for exponential families, with application to the Wisconsin Epidemiological Study of Diabetic Retinopathy. *The Annals of Statistics* **23** 1865–1895.

WALKER, S. (2004). Modern Bayesian asymptotics. *Statistical Science* **19** 111–117.

WALKER, S. and HJORT, N. L. (2001). On Bayesian consistency. *Journal of the Royal Statistical Society, Series B, Methodological* **63** 811–821.

WAND, M. P. (1994). Fast computation of multivariate kernel estimators. *Journal of Computational and Graphical Statistics* **3** 433–445.

WASSERMAN, L. (1998). Asymptotic properties of nonparametric Bayesian procedures. In *Practical Nonparametric and Semiparametric Bayesian Statistics*. Springer-Verlag. New York, NY.

WASSERMAN, L. (2000). Bayesian model selection and model averaging. *Journal of Mathematical Psychology* **44** 92–107.

WASSERMAN, L. (2004). *All of Statistics: A Concise Course in Statistical Inference*. Springer-Verlag. New York, NY.

WEISBERG, S. (1985). *Applied Linear Regression*. Wiley. New York, NY.

WONG, W. H. and SHEN, X. (1995). Probability inequalities for likelihood ratios and convergence rates of sieve MLEs. *The Annals of Statistics* **23** 339–362.

YU, K. and JONES, M. (2004). Likelihood-based local linear estimation of the conditional variance function. *Journal of the American Statistical Association* **99** 139–144.

ZHANG, C.-H. (2002). Risk bounds in isotonic regression. *The Annals of Statistics* **2** 528 – 555.

ZHANG, P. (1991). Variable selection in nonparametric regression with continuous covariates. *The Annals of Statistics* **19** 1869–1882.

ZHAO, L. H. (2000). Bayesian aspects of some nonparametric problems. *The Annals of Statistics* **28** 532–552.

List of Symbols

$\mathbb{R}$	real numbers
$\mathbb{P}(A)$	probability of event A
F_X	cumulative distribution function
f_X	probability density (or mass) function
$X \sim F$	X has distribution F
$X \sim f$	X has distribution with density f
$X \stackrel{d}{=} Y$	X and Y have the same distribution
IID	independent and identically distributed
$X_1, \ldots, X_n \sim F$	IID sample of size n from F
ϕ	standard Normal probability density
Φ	standard Normal distribution function
z_α	upper α quantile of $N(0,1)$: $z_\alpha = \Phi^{-1}(1 - \alpha)$
$\mathbb{E}(X) = \int x \, dF(x)$	expected value (mean) of random variable X
$\mathbb{V}(X)$	variance of random variable X
$\mathsf{Cov}(X, Y)$	covariance between X and Y
$\stackrel{P}{\longrightarrow}$	convergence in probability
$\rightsquigarrow$	convergence in distribution
$x_n = o(a_n)$	$x_n / a_n \to 0$
$x_n = O(a_n)$	$|x_n / a_n|$ is bounded for large n
$X_n = o_P(a_n)$	$X_n / a_n \stackrel{P}{\longrightarrow} 0$
$X_n = O_P(a_n)$	$|X_n / a_n|$ is bounded in probability for large n
$T(F)$	statistical functional (the mean, for example)
$\mathcal{L}_n(\theta)$	likelihood function

Distribution	PDF or probability function	mean	variance	MGF
Point mass at a	$I(x=a)$	a	0	e^{at}
Bernoulli(p)	$p^x(1-p)^{1-x}$	p	$p(1-p)$	$pe^t+(1-p)$
Binomial(n,p)	$\binom{n}{x}p^x(1-p)^{n-x}$	np	$np(1-p)$	$(pe^t+(1-p))^n$
Geometric(p)	$p(1-p)^{x-1}I(x\geq 1)$	$1/p$	$\frac{1-p}{p^2}$	$\frac{pe^t}{1-(1-p)e^t}$ $(t<-\log(1-p))$
Poisson(λ)	$\frac{\lambda^x e^{-\lambda}}{x!}$	λ	λ	$e^{\lambda(e^t-1)}$
Uniform(a,b)	$I(a<x<b)/(b-a)$	$\frac{a+b}{2}$	$\frac{(b-a)^2}{12}$	$\frac{e^{bt}-e^{at}}{(b-a)t}$
Normal(μ,σ^2)	$\frac{1}{\sigma\sqrt{2\pi}}e^{-(x-\mu)^2/(2\sigma^2)}$	μ	σ^2	$\exp\left\{\mu t+\frac{\sigma^2 t^2}{2}\right\}$
Exponential(β)	$\frac{e^{-x/\beta}}{\beta}$	β	β^2	$\frac{1}{1-\beta t}$ $(t<1/\beta)$
Gamma(α,β)	$\frac{x^{\alpha-1}e^{-x/\beta}}{\Gamma(\alpha)\beta^\alpha}$	$\alpha\beta$	$\alpha\beta^2$	$\left(\frac{\beta}{\beta-t}\right)^\alpha$ $(t<\beta)$
Beta(α,β)	$\frac{\Gamma(\alpha+\beta)}{\Gamma(\alpha)\Gamma(\beta)}x^{\alpha-1}(1-x)^{\beta-1}$	$\frac{\alpha}{\alpha+\beta}$	$\frac{\alpha\beta}{(\alpha+\beta)^2(\alpha+\beta+1)}$	$1+\sum_{k=1}^\infty\left(\prod_{r=0}^{k-1}\frac{\alpha+r}{\alpha+\beta+r}\right)\frac{t^k}{k}$
t_ν	$\frac{\Gamma\left(\frac{\nu+1}{2}\right)}{\Gamma\left(\frac{\nu}{2}\right)}\frac{1}{\left(1+\frac{x^2}{\nu}\right)^{(\nu+1)/2}}$	0 (if $\nu>1$)	$\frac{\nu}{\nu-2}$ (if $\nu>2$)	does not exist
χ^2_p	$\frac{1}{\Gamma(p/2)2^{p/2}}x^{(p/2)-1}e^{-x/2}$	p	$2p$	$\left(\frac{1}{1-2t}\right)^{p/2}$ $(t<1/2)$

Index

Springer Texts in Statistics *(continued from page ii)*

Madansky: Prescriptions for Working Statisticians

McPherson: Applying and Interpreting Statistics: A Comprehensive Guide,
Second Edition

Mueller: Basic Principles of Structural Equation Modeling: An Introduction
to LISREL and EQS

Nguyen and Rogers: Fundamentals of Mathematical Statistics: Volume I:
Probability for Statistics

Nguyen and Rogers: Fundamentals of Mathematical Statistics: Volume II:
Statistical Inference

Noether: Introduction to Statistics: The Nonparametric Way

Nolan and Speed: Stat Labs: Mathematical Statistics Through Applications

Peters: Counting for Something: Statistical Principles and Personalities

Pfeiffer: Probability for Applications

Pitman: Probability

Rawlings, Pantula and Dickey: Applied Regression Analysis

Robert: The Bayesian Choice: From Decision-Theoretic Foundations to
Computational Implementation, Second Edition

Robert and Casella: Monte Carlo Statistical Methods

Rose and Smith: Mathematical Statistics with *Mathematica*

Santner and Duffy: The Statistical Analysis of Discrete Data

Saville and Wood: Statistical Methods: The Geometric Approach

Sen and Srivastava: Regression Analysis: Theory, Methods, and Applications

Shao: Mathematical Statistics, Second Edition

Shorack: Probability for Statisticians

Shumway and Stoffer: Time Series Analysis and Its Applications

Simonoff: Analyzing Categorical Data

Terrell: Mathematical Statistics: A Unified Introduction

Timm: Applied Multivariate Analysis

Toutenburg: Statistical Analysis of Designed Experiments, Second Edition

Wasserman: All of Nonparametric Statistics

Wasserman: All of Statistics: A Concise Course in Statistical Inference

Whittle: Probability via Expectation, Fourth Edition

Zacks: Introduction to Reliability Analysis: Probability Models and Statistical
Methods

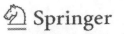 Springer
the language of science

springeronline.com

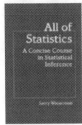

All of Statistics

L. Wasserman

This book is for people who want to learn probability and statistics quickly. It brings together many of the main ideas in modern statistics in one place. The book is suitable for students and researchers in statistics, computer science, data mining and machine learning. This book covers a much wider range of topics than a typical introductory text on mathematical statistics. It includes modern topics like nonparametric curve estimation, bootstrapping and classification, topics that are usually relegated to follow-up courses.

2004. 442 p. (Springer Texts in Statistics) Hardcover ISBN 0-387-40272-1

Inference in Hidden Markov Models

O. Cappé, E. Moulines, and T. Rydén

This book is a comprehensive treatment of inference for hidden Markov models, including both algorithms and statistical theory. Topics range from filtering and smoothing of the hidden Markov chain to parameter estimation, Bayesian methods and estimation of the number of states.In a unified way the book covers both models with finite state spaces, which allow for exact algorithms for filtering, estimation etc. and models with continuous state spaces (also called state-space models) requiring approximate simulation-based algorithms that are also described in detail.

2005. 664 p. (Springer Series in Statistics) Hardcover ISBN 0-387-40264-0

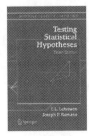

Testing Statistical Hypotheses
Third Edition

E.L. Lehmann and J.P. Romano

The third edition of *Testing Statistical Hypotheses* updates and expands upon the classic graduate text, emphasizing optimality theory for hypothesis testing and confidence sets. The principal additions include a rigorous treatment of large sample optimality, together with the requisite tools. In addition, an introduction to the theory of resampling methods such as the bootstrap is developed. The sections on multiple testing and goodness of fit testing are expanded. The text is suitable for Ph.D. students in statistics and includes over 300 new problems out of a total of more than 760.

2005. 786 p. (Springer Texts in Statistics) Hardcover ISBN 0-387-98864-5